Timber!

PAUL BRANNEN is Director of Public Affairs for the European woodworking and sawmill industries. He has written this book in a personal capacity, insisting on impartiality and retaining editorial control. His central argument (that a substantial increase in the use of timber in construction could significantly help tackle climate breakdown) predates his current role by many years. During his 15 years working on global development issues for Christian Aid, for instance, he focused on the impact of climate change on poverty in the developing world. While serving as a Member of the European Parliament and working on climate change legislation he became increasingly aware of the potential role for wood in decarbonizing the built environment.

Timber!

How wood can help save the world from climate breakdown

Paul Brannen

agenda
publishing

From my generation,
who should have done more,
for Poppy, Jack and Ella's generation
who will, alas, have to do more.

First published in 2024 by Agenda Publishing

Agenda Publishing Limited
PO Box 185
Newcastle upon Tyne
NE20 2DH

www.agendapub.com

ISBN 978-1-78821-735-4

British Library Cataloguing-in-Publication Data
A catalogue record for this book is available from the British Library

Typeset by Patty Rennie

Printed and bound in the UK by 4edge

If we are to achieve things never before accomplished we must employ methods never before attempted.

Francis Bacon (1561-1626)

Contents

Acknowledgements

Many people have helped me to write *Timber!* Without their assistance this book would never have made it to publication. Each chapter has had its facts and claims checked by both an academic and an industry specialist. I would like to thank my old school friend, Ian Peacock, who as an English graduate of Durham University was the obvious person to ask to check my prose, an exercise he carried out with great panache, and hence the second draft was infinitely better than the first. His consistently cheerful support throughout the writing process was much appreciated. Selfishly, I hereby claim all the mistakes as mine and mine alone.

In listing everyone else who helped I shall almost certainly miss one or two of you for which I apologise profusely. So, deep breath, here we go in reverse alphabetical order for the sole benefit of my old university friend Nigel Wright: Nigel Wright, Kris Wijnendaele, Craig White, Andrew Waugh, David Watson, Kristine Valujeva, Anthony Thistleton, Wil Srubar, Shaun Spiers, Matthew Rivers, Greg Richards, Ben Raskin, Michael Ramage, Clive Pinnington, Ian Paczek, Birgit Östman, Andrew Norton, Gary Newman, Constantin Muraru, Martin Milner, Silvia Melegari, Harald Mauser (to whom I sent more emails requesting information than to anyone else), Andy Leitch, Gerry Lawson, Charlie Law, Uwe Kies, Björn Källander, Adrian Joyce, Andy Howard, Dave Hopkins, Andrew Heald, Eleanor Harris, Robert Hairstans, Joe Giddings, Stuart Goodall, Mathias Fridholm, Jose Torero Cullen, Tony Colman, Paul Burgess, Jamie Bull, Ian Brown, Steve Briggs, Julie Bregulla, Helen Browning, Colin Brannen, Nick Boulton, Piotr Borkowski, Diego Benedetti, Jane Anderson and Lloyd Alter.

Finally, a specific thank you to Agenda Publishing: Camilla Erskine who has guided me through the process of writing my first book and Steven Gerrard who was of the view that there was indeed a book to be published on the subject of *Timber!*

Paul Brannen
Newcastle upon Tyne

Preface

I was lost. Hopelessly lost. Getting lost was an occupational hazard for a new Member of the European Parliament (MEP). The parliament building in Brussels is enormous. But help was at hand in the form of Frank Schwalba-Hoth, a former German Green MEP. He looked kind and was wearing a bowtie.

"Where are you trying to get to?" he asked.

I gave him the number of the committee room. "Right", he declared, "Follow me!" It was a fair walk and as we proceeded, "Have you decided?", he asked.

"Um, decided what?"

"Decided on your focus. You can't do everything. When you walk through parliament, you need people to point and say: 'That's Paul Brannen MEP. He's the Mother Teresa of . . . windfarms', or something along those lines".

We arrived at the committee room and with a shake of hands he was off.

It was a year before I saw him again. As it happens, it was at a climate change event I was speaking at, in the Chapelle pour l'Europe in Brussels. Herr Schwalba-Hoth was the chair. At the end, over a glass of Cabernet in the crypt, I asked him if he remembered our first meeting.

"Oh indeed".

"So, what am I the Mother Teresa of?"

"You, Paul, are the Mother Teresa of *trees*".

I've always had a few issues with Mother Teresa, so I mentally translated this into "the Martin Luther King of trees" which fitted in better with my theology.

From then on I was determined to become if not an expert at least reasonably well informed on trees, timber and their climate benefits (as opposed to becoming an elderly Albanian nun). Before long I was sitting on the European Parliament's Agriculture and Environment committees, focusing on forestry and on using more wood in buildings and the built environment.

Then Brexit struck. I was now a political Dodo. A "former British MEP". But, thanks to Herr Schwalba-Hoth I departed the Parliament with a niche understanding of European wood products and their potential to help decarbonize the built environment.

My next move was to become the Director of Public Affairs for the European Confederation of Woodworking Industries (CEI-Bois) and the European Organisation of the Sawmill Industry. An important part of my new mission was to represent the woodworking industries to the European Union institutions.

Since getting lost as a sapling MEP my knowledge of the wonderful world of wood has grown from acorn to oak. In this book I shall be advocating the climate benefits of using more wood in the built environment. *Timber!* is written not only for architects, foresters, builders and students but for all climate-conscious people who want to use wood intelligently to make a difference – especially campaigners and politicians at all levels of government. This book will, I hope, turn you into an advocate for the greater use of sustainable wood and wood-based materials, often in many surprising and counterintuitive ways.

It is not a silver bullet – there *is* no climate change silver bullet. However, I am convinced that more effective use of wood (currently misunderstood and misrepresented) is crucial to the fight to save the planet for our children and grandchildren. You do not need to be a lumberjack, tree hugger or ecologist to grasp this radical philosophy of wooden living.

My own journey into the woods, so to speak, has involved taking several roads less travelled. I am at heart a campaigner and my odyssey began in an unlikely place – a swamp-like football pitch in Leeds. As a student at Leeds University in the 1980s I spent a lot of time playing football, albeit it not very well. I played in goal for Clapham House where I lived in the first year. We had Wednesday afternoon fixtures against various departments such as chemistry, medicine (the medics were a vicious bunch) and law.[1] Matches were often postponed due to a waterlogged pitch. If this happened, I would join my friends Mike and Steve at the Students' Union Anti-Apartheid Movement (AAM) meetings where it was at least warm and dry.

The more it rained, the more I attended the AAM meetings. One particularly wet term I became a regular fixture and the Secretary asked if I would be interested in taking over from her the following year. Somewhat surprised at this request I answered, "Sorry Deb, I can't because, as you know, I normally play football on Wednesdays". Little did I realize that this was to be my "Road to Damascus" moment – a road which would eventually lead to an epiphany about wood in 2020 . . . and this book. Deb responded not so much snarkily as factually: "Oh well, if *that's* your priority . . .".

The arrow had been shot but it was several days before it was to hit its target. The following Wednesday it must have stopped raining as we had a match. We were having a hard time of it against the medics. I was in goal again and dreading having to dive into the muddy puddle looming ahead of me. I think we were losing 5–0 when I began wondering whether playing football badly in the mud was really more important than helping to end apartheid. Looking back, I am uncomfortable that I even felt there was a debate. But in my defence, an upbringing on Tyneside does rather come with added compulsory football. It was in my DNA. At the time it was a genuine dilemma.

In a foreshadowing of the politician I would eventually become, I reached a sensible, if irrational, compromise position with myself. I would not *seek* Deb out on campus. However if I *did* bump into her I would offer to take on the role. The matter was now in the hands of the gods. Being a theology student, I guess it was inevitable the gods would intervene

– and Deb and I did encounter each other a few days later in the Students' Union.

As a consequence, I ended my illustrious career as goalkeeper and took on the role of Secretary of the AAM, a role which then led me to work for the AAM in London after leaving university. I then did a stint as a Labour Party press officer in the ruthless machine forged by Peter Mandelson – an eye-opener on dealing with the media. Then I had another moment of epiphany while I was in the audience in London listening to Al Gore deliver his famous climate change presentation. I realized then that climate change would be at the centre of all my future endeavours. I subsequently became Christian Aid's first ever Head of Campaigns and Advocacy and made the controversial decision to concentrate on climate change. Christian Aid was the first UK development agency to do this. We focused on raising the issue with the UK's Protestant churches and the wider community. In 2007, we organized an ambitious "Cut the Carbon" march from Belfast to London, via Scotland and Wales and the Labour Party conference in Bournemouth. At 1,000 miles it was and remains the longest protest march in UK history.

In 2014, when I arrived as an MEP representing the North-East of England in the European Parliament, I was determined to continue working on climate change, which first entailed getting to grips with the policy-making process of the European Union. Let me lead you through the labyrinth. The process begins with a proposal drafted by the Commission and sent to the Parliament for debate and modification. Once the Parliament has voted for it, it moves into a "trialogue" lasting many hours and occasionally through the night. Eventually the Parliament, the Commission and the Council (chaired by a different member state every six months) come to a final legislative agreement. As well as binding legislation the process can also produce advisory policy which is not binding on member states but does influence them. Forestry matters have traditionally fallen into this category.

My arrival in the Parliament happened to coincide with a revision of EU Forestry Strategy and I was duly chosen by the Socialists and

Democrats, the second largest political grouping in the Parliament, which we Labour MEPs sat within, to be their "rapporteur". According to Wikipedia this is an "eminent role" with "substantial influence in the process leading to the adoption of EU-legislation". As a result, I found myself increasingly immersed in forestry policy and subject to the lobbying of all sorts of folk and organizations wanting to take me on visits to see different aspects of Europe's forests – in Finland, Romania, Belgium, the UK and France. By the time I came out the other side several months later I was well on the way to becoming, as predicted, "the Mother Teresa of trees".

I was then appointed as a rapporteur for a second time to help draft legislation on "Land Use, Land Use Change and Forestry". This led to another Eureka moment on my route to writing this book: *Planting trees is not enough to reduce carbon. We also have to chop them down . . . and use them in the built environment* (and then, of course, replace them with new planting). As a consequence, at my instigation, the legislation included text aimed at promoting a greater use of Long-Life Harvested Wood Products, including more use of wood in construction. So, when the curtain came down on UK membership of the EU, I did at least have a specific area of knowledge, thanks to Herr Schwalba-Hoth: forestry and the use of wood in construction.

This book draws on the knowledge and experience of forestry and the woodworking industries gained during my time in Brussels. It not only makes the case for timber construction but is also a call to action. In the battle against climate change we should all be worried, frustrated and angry, but we need solutions too if we are not to lose hope. This book is focused on solutions. I hope it will enable you to see the wood for the trees and I hope it will make you want to *act*.

Abbreviations

AAM	Anti-Apartheid Movement
BECCS	bioenergy with carbon capture and storage
BPIE	Buildings Performance Institute Europe
CCS	carbon capture and storage
CEI-Bois	European Confederation of Woodworking Industries
CLT	cross laminated timber
COP	Conference of the Parties
DEFRA	Department for Environment, Food and Rural Affairs
EPBD	Energy Performance of Buildings Directive
EPD	Environmental Product Declaration
EU	European Union
EURAF	European Agroforestry Federation
FTA	flame-retardant treatment
GHG	greenhouse gas
Glulam	glued laminated timber
IPCC	Intergovernmental Panel on Climate Change
JRC	Joint Research Centre (of the European Commission)
LCA	life cycle assessment
LVL	laminated veneered lumber
MMC	modern methods of construction
MtC	million tonnes of carbon
NAI	net annual increment
nZEBs	nearly zero energy buildings
RCW	ramial chipped wood
SRC	short rotation crop

1

What is the problem we are trying to solve?

Fact: *The building and construction sector is not on track to achieve decarbonization by 2050. And the gap between the actual climate performance of the sector and the decarbonization pathway is widening.*[1]

This is an optimistic book, but it starts from a challenging place.

In 2020 the global population was approximately 8 billion. By 2050 it will likely hit 10 billion.[2] The additional 2 billion people will all have something in common: they will need somewhere to live. The majority of these 2 billion people will be in the Global South, in urban areas, and will be housed in high-density, high-rise buildings.[3] They will need to cool their homes, mainly flats, not heat them. So next time someone tells you we need to build *less*, not more, ask them where they expect the 2 billion to live. Population growth may be flatlining in many European countries, raising for them the valid question: Do we really need to build more homes? But in the countries of the Global South, and in countries like the UK, USA and Canada, the population continues to rise – and we all need somewhere to live. It is this increasing demand for housing that puts humanity in such a challenging place. This population growth will create further massive demand for new buildings and the production of cement, steel and other building materials associated with this wave of construction will become a major source of greenhouse gas emissions (GHG).

To meet this construction demand the world will need to add 190 billion m^2 of new buildings every decade for the next 40 years, which is equivalent to adding a London-sized city to the planet every 70 days.[4] The challenge, as defined by carbon finance company Aureus Earth, is that, "Without intervention, business-as-usual construction would alone consume 60% of the remaining global carbon budget (500GtCO_2e) if humanity is to limit global temperature rise to 1.5°C".[5]

How we build, renovate, heat and cool our built environment has become the *biggest cause of climate breakdown* and is the central problem addressed in this book.

There are different ways in which our global carbon emissions can be subdivided, however using the sectors detailed in Table 1.1 is a standard approach.[6] To ascertain the total emissions from the built environment two of the five figures need to be combined: the operational emissions (predominately heating and/or cooling) at 28 per cent and the building materials and construction at 11 per cent. The resulting rounded 40 per cent figure makes the built environment – under this subdivision of emissions – the largest single emitter of carbon ahead of industry and transport. Clearly, reducing carbon emissions from the built environment is a key step which we must take in tackling climate breakdown.[7] But first, how do these emissions come about?

Global CO_2 emissions by sector	Percentage
Industry	30
Building operations	28
Transport	22
Building materials and construction	11
Other	9

Table 1.1 Global CO_2 emissions by sector

Source: New Buildings Institute.[8]

Most "built environment" emissions result from heating and cooling our buildings in the winter and summer respectively, with the demand for cooling rising in line with rising temperatures caused by global warming. (Yes, there is a destructive cycle developing). To do this we need energy and most of our energy is still derived from fossil fuels. So, heating and cooling is responsible for the burning of a vast amount of fossil fuels year after year. This fact is highlighted in the findings of the United Nation's (UN) Environment Programme 2022:

> despite a substantial increase in investment and success at a global level lowering the energy intensity of buildings, the sector's total energy consumption and CO_2 emissions increased in 2021 above pre-pandemic levels. Buildings energy demand increased by around 4% from 2020 to 135 EJ – the largest increase in the last 10 years. CO_2 emissions from buildings operations have reached an all-time high of around 10 $GtCO_2$, around a 5% increase from 2020 and 2% higher than the previous peak in 2019. *The buildings and construction sector is not on track to achieve decarbonization by 2050.* And the gap between the actual climate performance of the sector and the decarbonization pathway is widening.[9]

What we must do therefore is make our existing buildings more energy-efficient, primarily through better insulation.[10] This process is normally referred to as "retrofitting" in the UK and "renovating" on mainland Europe. But it is a mammoth task. The UK alone has 29 million homes and the vast majority need to be retrofitted. In fact, the UK has the "leakiest" buildings in Europe with poor levels of insulation and energy efficiency. This is because the UK has the *oldest and worst built* building stock in Europe.[11] To make matters worse even the UK's newbuilds are poorly insulated and also require retrofitting.[12]

Some European countries have relatively energy efficient homes with Germany and Norway being the best insulated in Europe.[13] However, countries such as France and Belgium are doing less well and the

EU, as part of its "Renovate Europe" push, has identified that a shocking three-quarters of all buildings, that is *190 million*, need to be made more energy efficient. Chapter 3 will explore the significant climate saving role of insulation.

So, after heating and cooling, what is the next biggest source of carbon emissions from the built environment? It is the creation and transportation of the *materials* we use to make our buildings and here a single construction material stands out head and shoulders above all others: concrete. As an article in *The Guardian* newspaper strikingly put it:

> In the time it takes you to read this sentence, the global building industry will have poured more than 19,000 bathtubs of concrete. By the time you are halfway through this article,[14] the volume would fill the Albert Hall and spill out into Hyde Park. In a day it would be almost the size of China's Three Gorges Dam. In a single year, there is enough to patio over every hill, dale, nook and cranny in England.[15]

Concrete is artificial rock. Its manufacture is responsible for 8 per cent – almost a *tenth* – of *all global carbon emissions*. If concrete were a country (what a thought), it would be the third biggest carbon emitter after China and the USA (see Table 1.2).[16] Carbon emissions associated with concrete are greater than those of the entire EU: 2.8 billion tonnes annually from concrete, as opposed to the EU's 2.7 billion tonnes.[17]

The emissions from concrete are inextricably linked with its key ingredient: cement. The manufacture of cement is problematic for two reasons both of which result in massive carbon dioxide (CO_2) emissions. First, making cement involves the process of extracting lime from limestone, triggering a chemical reaction which releases CO_2 into the atmosphere. Second, this reaction can only take place at the exceedingly high temperature of 1,400°C as the limestone is baked. The energy to do this is usually generated by burning fossil fuels resulting in yet more CO_2 being released into the atmosphere.[18] And the problem is exacerbated when concrete is strengthened with steel to make *reinforced* concrete. The manufacture of steel – smelting – also requires massive amounts of energy to achieve the necessary high temperatures, temperatures which electricity struggles to reach and hence the need to burn fossil fuels.[19]

Country	**Percentage of global carbon emissions emitted**
China	27
USA	15
Concrete	8
India	6.8
Russia	4.7
Japan	3.3

Table 1.2 Carbon emissions by country, including concrete

Source: Our World in Data.[20]

The global construction industry is therefore using two materials – concrete and steel – which have some of the highest levels of embodied carbon of any material on earth. It is called "embodied" carbon because concrete and steel simply would not exist without burning fossil fuels. The fossil fuels are the primary culprits, inserting their carbon footprint into the building materials. Over the past 150 years concrete and steel have become the two dominant materials used to construct our urban areas and city centres around the world. (We shall discuss embodied carbon further in Chapter 2).

Concrete is relatively cheap to produce and, in turn, purchase. As a consequence, huge volumes are often used indiscriminately rather than being deployed strategically for specific tasks for which it is best suited, such as creating foundations. Achieving such volumes requires vast amounts of sand extracted from beaches, rivers and seabeds. Sand, while a natural material, is not a renewable material and therefore, mining sand is not a sustainable activity. Across the world, in areas where it is needed, it is becoming an increasingly scarce resource and its mining disturbs aquatic ecosystems, compromising their ability to absorb carbon. [21]

Concrete and steel's dominance began with the iconic skyscrapers of American cities and the skyscraper pandemic continues today, especially in the Global South. Here rampant urbanization has driven high demand for tall, high-density home construction: skyscrapers, if you want them to sound glamorous; tower blocks if you do not. The high-rise has now migrated from the USA to China, which today uses almost half the world's concrete,[22] in fact China used more cement in the three years 2011–13 than the USA used in the whole of the twentieth century.[23] This forms part of a global trend in which the richer countries in the Global North have seen a levelling off or even a decline in their city-centre populations so are building fewer high-rise residential buildings (although not in London); but the cities in the Global South are still building upwards as well as outwards.

For a snapshot of city-centre housing demand in the future Egypt is a useful example. In 2022 the population of Egypt was recorded as 105 million (the UK's was 67 million) and is forecast to reach 124 million by 2050,[24] the year when the world has to reach net zero carbon emissions. The additional 19 million people will predominately live in the country's major urban centres led by Cairo with a current population of 21 million. Six and a half million, twice the population of Madrid, will live in a huge new administrative capital that is being built to the east of Cairo, on a swathe of desert equal to the size of Singapore.[25] This booming population is being housed in an ever-expanding network of high-rise tower blocks built from concrete and steel (see Figure 1.1). Cairo is the world's seventh largest city. Those above it are Mexico City and Sao Paulo (22 million), Dhaka (23 million), Shanghai (29 million), Delhi (33 million) and Tokyo (37 million).[26] All these cities, and hundreds of others, continue to expand, racking up carbon emissions via the construction of more buildings and then via their use (predominately cooling and heating).

Figure 1.1 Typical construction in Giza, Egypt's third largest city, using concrete and steel, August 2023

Source: iStock; Tamer Soliman.

A rising global population is nothing new, but 9.7 billion people by 2050 needs to be planned for.[27] For decades the question was: can we feed all these people? To which the answer has been "Yes", if we consume less, waste less and move away from meat and dairy-intensive diets. But now, we are faced with an even more challenging question: can we house all these people without destroying the climate on which we all depend? Having a place to live is a human right under the 1948 United Nations' Universal Declaration of Human Rights with Article 25 stating, "We all have the right to enough food, clothing, housing and healthcare for ourselves and our families".[28] As it stands these urban homes will be high-density and built predominantly from concrete and steel. Emissions from the built environment will, therefore, continue to rise, both as a percentage and in overall terms. Worse still, if other sectors are successful in reducing their emissions then the built environment's contribution will soar remorselessly towards 50 per cent of total emissions. We are building our own mausoleum – like the Egyptian pharaoh Ozymandias ("Look on my works, ye mighty, and despair!") – from the toxic twins of concrete and steel as the planet continues to heat up and climate breakdown looms.

Understandably, most people have little idea of how instrumental many of our everyday construction materials – especially concrete – are in causing climate change. These materials do not come with the equivalent of a health warning – unlike say a packet of cigarettes or a sugary drink. As ecological builders King and Magwood from North America note, "Just as we expect to know the mileage of a car or truck, the fat content of prepared food, or the presence of cancerous substances in anything we might use before purchase, we as a society, as a world in climate disruption, must know the climate impacts of what we use. Concrete, probably more than anything else, will be among the first to be subject to that expectation".[29]

Like all crises climate change has its weird nuances and attempts at spin. If you tune in closely to the debate about carbon emissions from the built environment you will hear talk of "green concrete" and "green steel". Time to choose my words carefully. It is not only about building materials, "green" or otherwise. It is about *proportion*. We need to do more with less – *whatever* the material. Increased material efficiency – building more with less – is of critical importance in our effort to save the planet. Reusing and recycling materials we have already made (having released CO_2 while making them) is something we have only just begun to grasp, and this requires lateral thinking but not necessarily new technology. We already have, for instance, new road surfaces partly made from recycled plastic.[30] However green we want to be we must reluctantly accept that concrete and steel have unique properties that make them ideal for use in certain constructions. So, we will inevitably go on using them. This book is not a naïve manifesto for banning the use of concrete and steel. However, when it comes to the built environment, there are many tasks where replacement materials are readily available – materials that have considerably lower levels of embodied carbon – that can perform just as well. Whichever way you look at this we are talking about using considerably less concrete and steel, even if it is misleadingly labelled as "green".

"Green concrete" is a ridiculous oxymoron. Yes, it is possible to make a type of concrete that has a lower level of carbon emissions than "traditional concrete". And yes, "green concrete" is a "less worse" product. But in the context of climate change and population growth, the gains are limited. Even if all future concrete was "green" and had an emissions rate that was 25 per cent lower than today's concrete, it would still be responsible for at least 6 per cent of all carbon emissions. However, even this "saving" would soon be wiped out because we are in the process of *doubling* the floor area of the world's buildings by 2060.[31] These "green" gains alone will simply not be enough to significantly cut the sector's emissions and we may find that a single material, increasingly labelled "green", could still be emitting more carbon by 2050 than the *whole of the EU put together*. Put another way, if someone you knew went down to the beach yesterday and emptied eight bags of rubbish all over the sand but today went down to the beach and removed *two* bags of rubbish, would you congratulate them for being "green"?

The evidence is . . . well, concrete. Concrete is a massive problem when it comes to climate change so we must use a lot less of it. This would still apply *even* if we reduced its carbon emissions during manufacture by 50 per cent. No one, even in the concrete industry, is claiming we are close to achieving this. But we will still need the stuff, so let us be pragmatic and reduce the emissions associated with its manufacture as much as possible. And let us stop calling it "green concrete" because it simply is not. But "less worse" is always better that "worst". Likewise with steel.

Steel does have one very big advantage over concrete in that it can be recycled almost indefinitely. In fact, it is one of the most recycled materials on the planet.[32] The industry claims that 81 per cent of steel goes on to be recycled.[33] However what they are rather quiet about is that to recycle steel you need to melt it down and to melt it you need heat, which typically involves the burning of fossil fuels producing . . . more CO_2. But, thankfully, unlike concrete the chemical process of making steel does not release significant amounts of CO_2. It is the need for heat that is the problem with steel, irrespective of whether you are making it new or recycling it. In the future, if we can make or recycle steel with lower levels of emissions, then this is to be welcomed and there is plenty of work underway on this.[34] We also need to recognize that we absolutely need steel to enable us to reach net zero, for instance to create offshore wind turbines[35] and electric vehicle charging points.[36] Meanwhile there is the opportunity to use existing steel in construction more efficiently as noted by the UK's Green Alliance given that "steel beams are often designed to use 30 per cent more steel than required and steel framed buildings are being built with twice the steel necessary to meet safety standards".[37] Timber frames are also often over-engineered in a similar materially wasteful way (see Chapter 5).

The big climate break through will come if we can make/recycle most steel with heat generated from electricity or hydrogen created from green energy (which could include biomass and nuclear) rather than from the burning of fossil fuels.[38] Where a traditional coal-fired blast furnace is replaced with an electric arc furnace energy use can be reduced by 60 per cent.[39] Europe, led by Sweden is endeavouring to do this but it is expensive. Europe's biggest steel-maker ArcelorMittal estimates it will cost up to £30 billion to decarbonize its operations.[40] Meanwhile many countries around the world are still building new coal-based furnace equipment. Europe produces only 8 per cent of the world's steel whereas China's steel industry produces over half, a manufacturing process powered by coal/coke that in turn is the cause of one-third of China's CO_2 emissions.[41] The global challenge is immense with one academic in the field Koch Blank observing:

> that all the steelworks operated by companies currently investing in, or committing to, zero-carbon technologies only represent about 8% of global steel production. Thus . . . a hundred-fold increase in the speed of transition is required in order to conform with the 1.5°C pathway agreed under the 2015 Paris Agreement. It would imply moving from converting one blast furnace coke oven every 10 years to ten furnaces every year.[42]

There is the option of deploying carbon capture and storage (CCS; see Chapter 8) in conjunction with steel manufacture, but as discussed in Chapter 8 this technology remains expensive and only partially developed for application at scale globally. In the meantime, let us be clear "green steel" is simply "less CO_2-intensive steel", "less embodied-energy steel", "less climate-bad steel".

To make matters worse the terrible twins, concrete and steel, have a bunch of mates with similar traits namely: brick, block and plasterboard. Admittedly, the unassuming house brick does not look like a climate disaster. The problem is that more than 1,500 billion bricks are produced each year globally.[43] Stacked end to end, the pile would reach to the moon and back 390 times. Bricks are formed from clay and then baked in an oven using a great deal of heat. We are once again burning fossil fuels. As noted by the UN Environment Programme, it has been estimated that China produced 800 million bricks in 1994, resulting in 43 million tonnes of CO_2 being emitted.[44] That is more than the annual CO_2 emissions of the Republic of Ireland.[45]

It is much the same with breeze blocks or cinder blocks, those "big grey bricks" which form internal walls and the inner exterior-facing walls but then disappear behind the more attractive bricks. They too require hot temperatures during manufacture, as does plasterboard with which we cover internal walls with before skimming it with plaster. Glass manufacture is also heat intensive.

Yet despite the global recognition that we must decarbonize the built environment if we are to have any chance of reaching net zero by 2050, we are currently going in the wrong direction. While building homes for 2 billion more people is something we must do so too is decarbonizing the built environment. There can be no solution to climate change unless we reduce the emissions associated with concrete and steel and their fellow carbon-intensive building materials.

That is where wood comes in, as an alternative building material. Did you know that there are urban skyscrapers in Norway, Austria, USA and Canada built from timber? Could *chopping down trees* be the solution to halting climate catastrophe? In this book I am advocating a radical change in approach to how we build. By using much more *wood* both to construct and renovate our built environment, we can actively reduce our carbon emissions.

Timber!

2

How timber can decarbonize the built environment

Fact: *per square metre buildings have the potential capacity to store more carbon than forests.*[1]

It is difficult to find people who do not like trees. Trees have cheerleaders across the social classes; young and old, rich and poor, left and right. There is a vocal minority who believe that trees should be left well alone. What, they ask, is the environmental gain of chopping them down and bringing them out of the forest? Well, the answer is that when it comes to tackling climate change our forests are only half the story – and hence only half the answer. We can build with wood, wood that *stores* carbon and *substitutes* for much more carbon-intensive building materials. Without doubt wood is special. As one of the champions of building with wood, Vancouver-based architect Michael Green, has put it: "Wood is the most technologically advanced material I can build with. It just happens to be that Mother Nature holds the patent on it and we are not comfortable with that. But that's the way it should be: nature's fingerprints in the built environment".[2]

We have been building houses from wood for thousands of years. Most European cities still have a few timber-framed buildings dating back 300 years or more. York in the UK is a good example. Here you will find The Shambles – a street of medieval timber-framed buildings with a strong Harry Potter feel to them. Strasbourg, home of one of the two seats of the European Parliament – and somewhere I travelled to every month for five years as an MEP – has some fine examples of old timber-framed buildings (Figure 2.1). Some of these were carefully rebuilt after the Second World War but to the untrained eye look as old as those that were not damaged.

Figure 2.1 Old timber-framed buildings in Strasbourg, France

Source: Philippe de Rexel.

When these buildings were erected, they had only one purpose – that of providing shelter, a home. However, although unknown at the time, they were also safely storing the carbon that the timber had sequestrated (soaked up) when it was growing as a tree in the forest. As you may remember from your biology lessons, trees absorb carbon dioxide (CO_2) and release oxygen (O_2) – holding onto, storing, the carbon that was in the carbon dioxide. For every dry tonne of manufactured timber around 1.8 tonnes of CO_2 is removed from the atmosphere.[3] Chopping the tree down *does not* release the carbon, it remains stored in the wood. If we burn the wood, we *do* release the carbon into the atmosphere, where it joins up with oxygen to form CO_2 – the very gas whose effects we are trying to mitigate, as its relentless increase is causing the climate to change. If we build with the wood, we safely store the carbon in the built environment – potentially for hundreds of years. This is a good thing to do. At the same time, back in the forest, we replant the felled trees with new saplings – often three saplings per felled tree – to ensure the process is circular and hence sustainable. As noted by Himes and Busby, "In sustainably managed forests, new trees are regenerated to replace trees that are harvested so that there is no net loss of forest carbon".[4]

Today's climate crisis demands two responses. First, we need to *reduce*, and wherever possible to stop generating, further CO_2 which is by far the most important action we can take. Secondly, we also need to *remove* a colossal amount of CO_2 from the atmosphere and store it safely. Trees and the timber they produce are fantastic at storing carbon.

We know how much carbon Europe's forests store. If we add together the carbon in the above-ground biomass (what we can see of the tree) and the carbon in the below-ground biomass (the roots) then we have 13,240 million tonnes of carbon (MtC) (13.24 gigatonnes (Gt)).[5] However we do not have any research findings or official figures for how much carbon is stored in all the wood in the European built environment. This is an omission that needs to be remedied. How can we measure an increase in the amount of carbon stored in the built environment – something we need to do – if we do not have a baseline to work with? A conservative *estimate*, made by someone with knowledge in this field, is that we *may* have 1,800MtC (1.8Gt) of carbon stored in the timber products contained within the European built environment,[6] which would be twice as much carbon as stored in the Swedish forest (1,000MtC (1Gt))[7] and seven times more than the UK forest (250MtC (0.25Gt)).[8] These are estimates and it needs reiterating that we do not know how much carbon is currently stored in the timber deployed in the European built environment. We need to know.[9]

When a tree is felled, it can no longer absorb carbon but the wood continues to hold, or store, the carbon it has previously sequestered. However, it should be noted that a growing tree cannot go on absorbing carbon forever. Older trees start to slow down on absorption before stopping altogether and dying, whereupon the rotting trees begin to release their carbon back into the atmosphere as methane (CH_4), which is also a greenhouse gas, one 80 times more potent than CO_2, albeit with a shorter lifespan (see Figure 2.2).

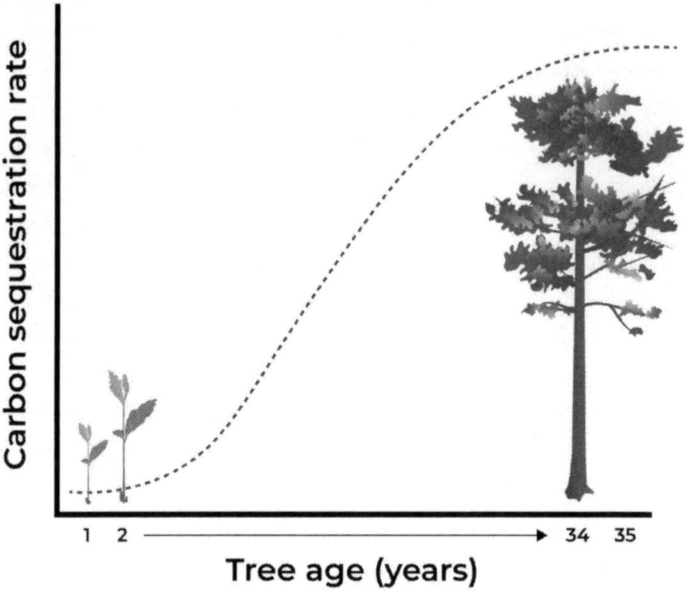

Figure 2.2 How the carbon capture capacity of trees change as they age

Source: Carbon Neutral, Australia.[10]

While it may seem a shame to fell a tree in its prime, from a climate perspective this is the ideal time to do so – before growth and hence carbon absorption slows and stops. (Farmers do the same as foresters – they harvest their crops at the optimum point in their life cycle.) We can then move the maximum amount of carbon stored in the timber into wooden products in the built environment, for example, house frames, floorboards, beams and roof joists.

Given scientists across the world are (rightly) working hard to ascertain how we can mechanically/chemically remove carbon from the atmosphere and then store it safely underground (a process known as carbon capture and storage – CCS), why would we not also want to maximize the potential of the built environment to safely store as much carbon as possible? Every existing building in Europe could safely store between 2–5 tonnes of carbon in wood products including wood fibre insulation (see Chapter 3.)

That we currently do not prioritize the storing of carbon in wood in the built environment given we are facing climate breakdown is a major mistake. That there are vested interests that would be harmed by an increased use of wood in construction should come as no surprise, but equally many people actively working to tackle climate change have missed or greatly underestimated the power of wood in the race to decarbonize the built environment. These include John Schellnhuber, who has driven climate change understanding in Germany, Europe and beyond since setting up the Potsdam Institute of Climate Impact Research in 1992. As a longstanding member of the UN's Intergovernmental Panel on Climate Change (IPCC) his work has influenced both Angela Merkel (the former Chancellor of Germany) and Ursula von der Leyen (the current President of the European Commission).

Schellnhuber has acknowledged that he "missed" wood's potential to store carbon in the built environment until he was well into his 60s, years after he began his work on climate change. In a way it is reassuring that even this eminent and visionary professor was initially unable to "see the wood for the trees". His more recent academic work, in particular his contribution to the often-cited Churkina *et al* paper "Buildings as a global carbon sink", has promoted the idea of the "forest pump".[11] Put simply, we grow trees in the forest and they sequester carbon. We then fell the trees. We turn the trees into wooden building products. These products store carbon in the built environment for 50, 75, 100 years. Meanwhile we replant new trees, saplings, and they start the carbon sequestration process all over again, and will in time produce more timber for use in the built environment, and so on. In the words of Schellnhuber:

> This is the motto: we have to reforest the planet and retimber the city. To restore the climate, we need to, first, add and support 500 billion trees (that is 50 trees per every human being), and second, build 2 billion homes from harvested biomass. This is the nature-based solution. [. . .] Nature did a fantastic job of extracting CO_2 from the atmosphere: we need to replay this process, by turning cities into organic mass.[12]

That his work has reached the top of the European Commission was evidenced when in September 2020, in her first State of the Union address to the European Parliament, President von der Leyen said: "Our buildings generate 40 per cent of our emissions. They need to become less wasteful, less expensive and more sustainable. And we know that the construction sector can even be turned from a carbon source into a carbon sink, if organic building materials like wood and smart technologies like AI are applied".[13] This observation when publicly and surprisingly delivered on the European stage was well received by those of us who inhabit the world of wood. There is no question that the built environment's ability to store carbon is massive. For instance, a 2021 report from ASN Bank and Climate Cleanup stated that there was the potential in the Netherlands to use the projected one million new homes required before 2030 to store 50MtC. This is a quarter of the country's annual emissions.[14]

The USA is also waking up to wood's carbon-storing ability. A 2023 report from the Rocky Mountain Institute recognized that the construction of new homes in the USA resulted in 50MtC of emissions annually.[15] However, it went on to argue that the very same homes could store significantly more carbon than their construction generated if carbon-storing materials were deployed in their construction, helping the US reach its climate targets quickly and efficiently. The Institute referred to this opportunity as "low-hanging fruit".

Timber's ability safely to *store* carbon in the built environment is therefore a great asset in the battle against climate change but the good wood news does not end there because wood has another valuable attribute, the ability to *substitute* for other materials that are significantly more carbon intensive. This may seem a statement of the obvious – when we build with timber, we do not use concrete – but it is often missed when carbon emissions are calculated for building work.

In their insightful report *Why we need more trees in the UK*, Friends of the Earth spell out this important contribution that timber can make: "We should also be increasing timber use for house building and construction, to reduce the use of high-carbon materials such as cement, mortar and brick".[16] If you build your block of flats out of wood rather than concrete you will *replace* the concrete, a material that was made by burning fossil fuels, with a material that is essentially the opposite. Wood is a safe store of carbon that has been taken out of the atmosphere, not a material whose manufacture has put more CO_2 into the atmosphere. Recent studies on the impact of substitution suggest that on average one tonne of CO_2 stored in wood represents 1.2 tonnes of CO_2 avoided through the replacement of conventional materials.[17] This is the "low-hanging fruit" referred to in the Rocky Mountain Institute report.

We need to be careful here with the terminology as it can get rather confusing. Wood advocates and climate specialists talk about "embodied carbon" in relation to building materials, a term which sounds similar to "stored carbon" or very similar to "embedded carbon". However, the first needs to be reduced as much as possible, the second increased as much as possible and the third term is probably not being accurately used.

"Embodied carbon" is a key term in understanding how wood can help save the world from climate breakdown. The term refers to the greenhouse gas emissions resulting from the manufacture, transportation, installation, maintenance and final end use of building materials. It describes the cradle-to-grave carbon footprint of a building material. The term is often juxtaposed with "operational carbon". Essentially embodied carbon is generated during the construction process whereas operational carbon is generated when the building is in use – primarily heating and cooling but also lighting (see Table 1.1).

All building materials, to varying degrees, are responsible for embodied carbon because CO_2 was released into the atmosphere as part of the manufacturing process, usually because fossil fuels were burnt in the process – hence the use of the word "embodied" as it is intrinsic to the process. A small number of building materials may have been manufactured in such a way that they have virtually no embodied carbon but they will need to be transported at some point and this is likely to involve fossil-fuel powered transport. Concrete and steel both have very high levels of embodied carbon, as we noted in Chapter 1, as it is in their DNA. Wood is the opposite. Yes, wood has a relatively small amount of embodied carbon as a result of being transported from the forest, cut in the sawmill, then transported to the construction site, i.e. fossil fuels were burnt in the process of turning it from a tree into a timber roof beam for a house. However, the roof beam could have been concrete and the beneficial figure from a climate perspective is the difference between the CO_2 emitted to make a wooden beam and the CO_2 emitted to make a concrete beam. This figure, this difference, is a measure of the *substitution effect – a figure as valuable in our endeavours to get to net zero as the storage figure.* Scale this up from city to city, country to country, continent to continent and we can drive down carbon emissions from the built environment via this substitution role – yet more low-hanging fruit.

Calculations made by Michael Ramage at the University of Cambridge have shown that erecting a 300m^2 four-storey wooden student residence in the city generated 126 tonnes of CO_2 emissions. If it had been made with concrete the tally would rise to 310 tonnes. If steel had been used emissions would have topped 498 tonnes. Indeed, from one point of view, the building might actually be viewed as "carbon negative" as there is the equivalent of 540 tonnes of CO_2 stored in the wood resulting in a long-term subtraction of CO_2 from the atmosphere.[18]

The Potsdam Institute for Climate Impact Research have similar examples that they deploy. They also make an interesting observation with regards to land use. Trees need land to grow on and buildings need land to be constructed on. Per square metre, which can potentially store more carbon, a plot of forest or a plot of land with a timber building? The answer according to the Institute is the latter: "A five-storey residential building structured in laminated timber can store up to 186kgC m^{-2} in the primary structure, that is more than in the aboveground biomass of the natural forest with the highest carbon density (52kgC m^{-2}, typical for the Coast Range ecoregion of North America)".[19]

Please note, generally speaking wood has a very *small* amount of embodied carbon in comparison to the majority of building materials. It is wood's low embodied carbon that helps it to deliver a big positive climate impact.[20] However specific wood products such as wood fibre insulation can have a higher level of embodied carbon than you might expect, as we shall discover in Chapter 3.

"Stored carbon" is what it says, carbon stored in a product, in this case in a wooden product and the longer it stays there the better for the climate. Greensted Church in Essex, England, has a nave constructed from large split oak tree trunks that have been dated to between 1063 and 1108 – that is over 1,000 years of carbon storage. Westminster Hall, part of the Palace of Westminster, has an oak hammer-beam roof commissioned in 1393 – over 600 years of carbon storage.[21] As Churkina and his colleagues have documented: "Buildings, which are designed to stay for decades, are an overlooked opportunity for long-term storage of carbon, because the most-widely used construction materials such as steel and concrete hardly store any carbon".[22] If the UK were to build 300,000 new homes per year with timber frames, this could sequester the annual carbon emissions of a typical coal-fired power station.[23]

When people use the term "embedded carbon" what they actually mean in most cases is embodied carbon. However, it is sometimes used to mean stored carbon, i.e. the opposite of embodied carbon. Consequently, if you come across the term "embedded carbon" you will probably need to check exactly what the user is trying to say.

To sum up, we have two issues at play and they are both equally important: (1) wood's ability to continue storing the carbon that the tree sequestered in the forest as timber in the built environment; and (2) wood's ability to prevent the carbon emissions which *would* have happened if we had used concrete and steel, brick and block, instead of timber – wood's substitution impact. Or as the researchers at the Potsdam Institute have put it: "By employing bio-based materials, technologies and construction assemblies with high carbon storage capacity and low embodied emissions, we can create a durable human-made global carbon pool while simultaneously reducing CO_2 emissions associated with building sector activities".[24]

Much thought is currently going into how we can best incentivize these two climate beneficial properties of wood: storage and substitution. The obvious first step is to draw back the curtains and let the sunshine in, i.e. make this all much more visible to politicians, policy-makers, those who commission new buildings and the wider public. Without knowing what is going on, people cannot change things. This includes circumventing the vested interests who would prefer to keep the curtain closed on embodied carbon and the opportunity to safely store carbon in our homes.

The House of Commons in the UK may not on first consideration appear to be the place to turn to for enlightenment but to give them their due the members of the House of Commons Environmental Audit Committee published a veritable sunshine report on 11 May 2022 titled *Building to Net Zero: Costing Carbon in Construction.* To prepare their report they read 156 written submissions and held four public evidence sessions, hearing from 26 witnesses including academics, architects, developers, builders, professional bodies, construction and fire risk consultants, financial institutions, trade associations, government ministers and advisors. The Committee's number one conclusion was indeed a ray of sunlight: ". . . the single most significant policy the Government could introduce is a mandatory requirement to undertake whole-life carbon assessments for buildings".[25]

"Whole life", if interpreted properly, would mean an assessment of *both* the CO_2 resulting from making the materials to construct the building *and* the CO_2 emitted as a result of heating and cooling the building for the whole of its life. In the vast majority of cases the figures revealed by such assessments will be enough to scare any thinking person concerned for the future of our planet. It is only when the scale of the problem is fully visible that we can really get to grips with the solution. As Ramage has observed:

> Legislating to make it mandatory to measure and publicly declare the amount of embodied carbon in a new building or substantial renovation along with its predicted operational emissions is a crucially important first step – although one we should have taken years ago. However the climate crisis is now so pressing that the next step of legislating to set carbon emission limits on both construction and operation of all buildings must also happen now. Yes, we have to count the carbon and we also have to cut the carbon.[26]

Across the Channel our fellow Europeans are also waking up to wood's potential as a substitute and store of carbon. "EU to start measuring 'embodied' carbon emissions from buildings" reported Euractiv in November 2021.[27] One of the mechanisms for doing this is the EU's Energy Performance of Buildings Directive (EPBD) which found itself being revised before the environmentally friendly ink had dried on the first version such is the urgency to address this issue. The revised EPBD aims to set out how the EU member states can achieve a zero-emissions and fully decarbonized building stock by 2050. This is not so much about new buildings rather it is about the existing building stock, 75 per cent of which needs to be renovated (have its insulation improved) to such a high degree that the buildings become carbon neutral to run. This can be achieved if we drive down their energy use to as low a level as possible and then source the little energy they need from a green energy source.

The EU aims to phase out the use of fossil fuels in heating by 2040. In this context the EPBD will address carbon emissions over the full lifecycle of a building, through mandatory calculation and disclosure of this information for new construction, to inform citizens and business and raise awareness – what the House of Commons Environmental Audit Committee is asking for. Making good choices about efficient building practices and materials can have a huge effect on both operational and embodied carbon emissions. This is why this proposal is not only about energy efficiency and operational carbon emissions, but it is also the start of a forward-looking approach to match energy performance with resource efficiency, circularity and sustainability.

While I was an MEP I became much exercised by considerations surrounding how we could incentivize an increase in the use of timber in construction. Clearly *storing* carbon in timber in the built environment is a good thing, as too is *substituting* timber for more carbon-intensive alternatives, but how to do it in the construction industry – a sector which is known for its conservatism?

One idea that has ebbed and flowed is the contentious area of carbon offsetting. It is not to everyone's liking and it is a concept that many environmentalists oppose. I think it is understandable that some see the sale of carbon offsets as akin to the sale of indulgences in the Middle Ages. Essentially you did something bad and you mitigated this by buying from the church an indulgence to offset the sins you had committed. Both a money-spinner for the church and popular with the guilty masses. In a similar vein you pump out carbon into the atmosphere as a consequence of a long-haul flight or your business emits carbon as a result of its activities (which, all businesses do to a larger or lesser extent), but given that you are of a decent disposition you decide to mitigate these carbon emissions by offsetting them by doing some good. Good comes in the form of carbon credits that you can buy to offset your "bad climate behaviour".

Make a quick tour of the internet and you will be spoilt for choice on the offsets available for purchase. Popular offset sales include planting trees (to soak up carbon), supporting solar panels or wind turbines (to generate clean electricity), restoring a deep peat bog (which will soak up a surprisingly large amount of carbon), biochar[28], waste-paper insulation, the list goes on. If you can buy a credit issued against storing carbon in a newly planted forest, why is it that you cannot buy a credit issued against storing carbon in a newly built timber house? Actually, you can, but the choice is limited to only one company in the world who currently facilitates the sale of such credits. It is a US company called Aureus Earth, based in Texas.[29] In 2022 they made their first offset sale having quantified all the carbon stored in the timber used to build a new Founders' Building for the University of Washington (see Figure 2.3). The tonnes of safely-stored carbon were then monetized and sold for $150,000.[30] As it happens they were bought by benefactors of the university (they could have been bought by a pension fund, bank, a company) who then gifted them to the university. Presumably the University of Washington will use them to offset some of their carbon emissions and we can look forward to this being written up as a positive news story in an annual report.

Figure 2.3 The new Founder's Building at the University of Washington
Source: Tim Griffith.

Commenting on this world first co-founder of Aureus Earth Wil Srubar explained:

> Decarbonizing the built environment is a crucial component to effectively combating climate change. Mass timber presents an opportunity to turn new buildings into effective carbon sinks. By issuing offsets for the biogenic carbon stored in a mass timber building, we are helping developers offset the cost of mass timber construction, thereby accelerating the pace of the construction industry's transition to a low-carbon economy.[31]

Note Srubar's final point, this new income stream generated by the carbon offset sale can be used essentially to lower the cost of building with wood. Srubar also refers to timber with the prefix "mass", quite what this is and how it differs from ordinary timber will be explained in Chapter 4.

The work of Aureus Earth signals the early stirrings in the woodshed of something big and potentially game-changing on both sides of the North Atlantic, not least because they are keen to bring their approach of monetizing stored carbon to Europe. King and Magwood argue that in North America companies like Aureus Earth "are bringing much-needed financial instruments to incentivize the decarbonization of the construction industry and store biogenic carbon for generations".[32] In so doing, they are making carbon "a physical asset that can be monetized in new building and infrastructure projects".[33] This stored carbon offsets present-day emissions and is slowing the build-up of emissions, thereby buying us some much-needed time in the battle against climate change.

In April 2024 the European Parliament passed into law by a large majority the "Carbon Removal Certification Framework".[34] This will seek to provide a framework that will allow a wide series of initiatives to flourish across farming, forestry, industry and the built environment, which will incentivize the safe storage of carbon. In the case of the built environment this legislation will help facilitate carbon offsets issued against CO_2 stored in wood products. There is a distinct sense that the "plates are moving" and a "wall of money" stands ready to "invest" in these new carbon offsets. Analysts at BloombergNEF have predicted the price of offsets could rise significantly, creating a $550 billion market by 2050.[35]

Why so much money? Well, let us look ahead a little. As we close in on 2050 and the need to have reached net zero, companies, institutions and households will have to be carbon neutral. Currently such an approach is voluntary but as the years tick down it will in all likelihood become the law in the EU, North America, the UK and other countries. Hence the more progressively minded are getting ahead of the curve, sometimes due to shareholder pressure, with the aim of being able to table annual carbon neutral accounts. The priority remains to drive down their direct and indirect carbon emissions, but whatever remains outstanding is then negated by the purchase of carbon offsets.

Already, without legal compulsion, demand for carbon offsets is outpacing supply, especially for good quality offsets, sometimes referred to as "gold standard". With such demand, it is perhaps not surprising that some carbon offset schemes are not worth the recycled paper they purport to be written on and claims of "greenwashing" are often, alas, correct.[36] This is a barrel of apples that is already more than half rotten, but there are plenty of genuine, at least bronze- or silver-standard, schemes out there which are credible and do what they say on the tin. Hence in the EU thought is being given to devising a methodology that would enable timber buildings or renovations – with wood or insulation with wood fibre – to generate carbon credits for sale on the open market.

Views vary on who should be the initial recipient of the offsets: the customer, the building company, the architect, the wood product manufacturer, the sawmill, the forester? We shall have to wait and see, but any of these possible recipients would likely trigger a drop in the price of building in wood. If the manufacturers of timber house-frames were able to sell carbon offsets linked to their product they could then use this new income stream to lower the price of their timber frames. A reduction in the price of their timber frames could generate new customers. A further increase in sales would mean the generation of more offsets which in turn would mean more income triggering another price drop. Hence a positive process is unleashed resulting in a steady increase in more carbon being stored in the built environment directly due to a growth in the sale of carbon offsets.

The whole process would clearly need rigorous independent auditing to prevent potential fraud. The issuing of offsets, say, by the wood product manufacturer, needs to be audited to ensure the agreed methodology is being met. Likewise, the company buying the offsets to "offset their emission's gap" also needs to be audited to check their claims. Whether such carbon offsets should be available for anyone to buy is a bigger question. Do we really want cash-rich fossil fuel companies buying up vast quantities of offsets (and they can afford to) which then enables them to claim they are carbon neutral? Surely not, otherwise we are back with the moral equivalent of the medieval indulgence. As Friends of the Earth have noted: "The offset market is booming. Offsetting can sometimes be the genuine effort of a business or an individual trying to mitigate the impact of their residual carbon emissions (carbon emissions they can't get rid of). But mostly carbon offsetting is seen as a cheap way to carry on business as usual".[37]

What then of buying offsets with an eye to an investment? The drive to net zero will inevitably see the price of stored carbon rise steadily if not drastically in the next few years – that BloombergNEF figure could well be correct. A £1,000 of carbon offsets in wood products could be worth £3,000 in ten years' time, a handsome return. As a democratic socialist this is a manifestation of capitalism that I feel able to endorse, for without question we need to accelerate timber's role to *sequester*, *store* and *substitute* carbon – the three "S"s – in order to make a highly significant contribution to tackling climate change.

3

Insulation: the big climate win we have yet to deliver

Fact: *Wood fibre is highly sustainable, locking up considerably more CO_2 than is produced in its manufacture. It can therefore play an essential part in a strategy for mitigating climate change.*[1]

Loft insulation may seem an odd subject for a call to arms. But your loft, possibly your least exciting room, is critical to tackling climate change. That is because efficient insulation helps us use considerably less energy heating homes, which in turn means less fossil fuels will be burnt. But your current insulation material, if you have any, will most probably have both contributed to climate change as well as help prevent it.

The market for building insulation is already huge, worth £1.38 billion in the UK alone, and it will become even bigger as we endeavour to achieve net zero emissions from the built environment by 2050.[2] This is because both the UK and the EU are way behind targets for retrofitting existing building stock with a high standard of insulation.[3] Additionally, in most cases, our new builds also fall well short of the level of insulation necessary to help us achieve net zero and will themselves need to be (expensively) retrofitted with better insulation.

Most household energy use goes on heating our homes during the colder months (see Table 3.1).[4] The quickest, most effective and cheapest way of reducing this consumption of energy is to better insulate our homes. That means the warm air we have expensively created for ourselves does not simply leak away, creating a need for yet *more* warm air, which means burning yet more gas, oil or coal. Our homes are no different to our bodies. Feeling cold? Put a jumper on. House feeling cold? Insulate your loft and walls. Homes in the UK are on average older than homes in the rest of Europe and therefore tend to have poorer insulation. So, they need more heating and use more energy. Shockingly it is not only *old* homes that are badly insulated. *New* homes are as well.

Energy use	Percentage
Space heating	64.4
Water heating	14.5
Lighting and electrical appliances	13.6
Cooking	6
Other uses	1.1
Space cooling	0.5

Table 3.1 Energy consumption in EU households 2021

Source: Eurostat.[5]

Since the beginning of 2021 all new buildings constructed in the EU must be nearly zero energy buildings (nZEBs) under national laws that transposed the Energy Performance of Buildings Directive (EPBD). It is clearly early days for this policy but the Buildings Performance Institute Europe (BPIE) has found "a wide degree of disparity across Member States in terms of the implementation".[6] The BPIE also notes that the high standard set by nZEBs, which by necessity will involve a very high level of insulation, should in turn become a benchmark for the renovation of the existing building stock to a similarly high standard. In turn, the BPIE argues that the very low amount of energy needed for such renovated buildings should be both renewable and produced on-site or locally.[7]

The last Labour government in the UK took a step towards remedying the poor level of home insulation when Chancellor Gordon Brown pledged in 2006 "to ensure that within ten years every new home will be a zero-carbon home" and "to accelerate the building of zero carbon homes. For a time-limited period, the vast majority of new zero carbon homes will be exempted from stamp duty".[8] Needless to say, this was warmly welcomed by environmental groups. The plan was to phase in the zero-carbon target for new homes, however, the majority of house builders set about resisting the imposition of compulsory targets. Their opposition paid off when the incoming Conservative/Liberal Democrat government *scrapped* the zero-carbon homes policy.[9] This allowed Britain's house builders to go on recklessly building poorly insulated homes. Lord Deben (John Selwyn Gummer) Chair of the UK's Climate Change Committee 2012–23 was scathing towards the UK's major house builders: "We have built a million and a half homes which are not fit for the future, all of which will have to be retrofitted as a result. And the house builders have done a terrible thing, which is to take a profit today and hand a bill on to everybody who has bought one of those houses".[10] He is right. It is a disgrace that a family receiving the keys to an expensive new build still need to carry out costly work to get proper insulation. And, of course, it is harder and more expensive to retrofit an existing home, new or old. It is cheaper and easier to build the insulation into the home while it is being constructed.

While you may disagree with the tactic of blocking major roads as used by the campaign group "Insulate Britain", an offshoot of Extinction Rebellion, their climate message is spot on – we need to insulate Britain, the rest of Europe and North America *fast*. Currently we simply are not installing the insulation that we desperately need. In the UK 8 million lofts and 7.24 million solid walls need insulating.[11] That insulation rates are running way behind target was identified by the independent Climate Change Committee in their March 2022 report "Independent Assessment: The UK's Heat and Buildings Strategy": "Installation rates for building insulation have plummeted over the last decade, and are far below the level they need to be. Insulating homes is important not just for reducing emissions, but also to protect people from the impact of volatile energy prices"[12] (see Table 3.2).

Year	Number of lofts insulated
2010	1,180,000
2011	1,100,000
2012	1,600,000
2013	130,000
2014	200,000
2015	100,000
2016	60,000
2017	30,000
2018	40,000
2019	20,000
2020	30,000

Table 3.2 Installation of loft insulation in the UK
Source: Climate Change Committee.[13]

The Chair of the UK's National Infrastructure Commission, Sir John Armitt, was equally clear:

> A decade ago, we were seeing more than 2.3 million insulation measures a year, like lofts or cavity walls being insulated. Last year [2021] the figure for comparable schemes was less than 100,000. To keep on target for reducing carbon emissions as well as bills, we need attractive and accessible programmes that help homeowners with their lagging and double glazing. The bottom line is that

> the current state of the UK property stock means that a significant proportion of the ballooning bills we face is wafting out of windows and roofs across the land. A national effort to insulate and cut waste make sense for both our pockets and the planet.[14]

The huge drop off in loft insulations in 2013 coincides with the then-Prime Minister David Cameron apparently telling aides that the time had come to "get rid of the green crap".[15] The number of homes having their lofts or cavity walls insulated each year dropped by 92 per cent and 74 per cent in 2013 respectively and has never recovered. It is cold comfort that the EU member states are doing no better. A 2019 report produced for the European Commission found that most existing buildings were constructed with no regard to energy performance.[16] Despite the opportunity to make them more energy efficient, energy renovation rates are dire in all member states.

Some of the UK's existing 29 million homes will likely be demolished by 2050 but it is estimated that around 26 million (90 per cent of current stock) will still be standing. These will all have to be retrofitted to a high standard of insulation, i.e. generating minimal CO_2 emissions. If not, they will continue to contribute around 15 per cent of the UK's total carbon emissions.[17] At the current rate of retrofitting, it will take us 80 years to insulate them all properly and we would miss the net zero target of 2050 by *more than 50 years.*

People on low incomes are most adversely impacted by the lack of insulation in the housing stock, as noted by Adam Scorer, Chief Executive of National Energy Action. Most vulnerable to high energy prices and more likely to be living in the least energy efficient homes, low-income families would benefit the most. Scorer points out that despite legal commitments to bring all such homes up to a minimum energy standard by 2030, the evidence suggests the rate of progress is so slow that it will take more than *150 years.*[18]

To hit the 2050 net zero target, the retrofitting rate in the UK needs to be achieving 4 per cent per year, which is four times the current rate:

1 million homes per year, not a miserly 250,000. Reducing household fossil fuel consumption is fundamental to stopping runaway climate change. Switching to electric heating generated from green sources will help us transition away from fossil fuels but the overall energy demand in a world with a growing population means driving down energy use – irrespective of how we make it – is essential. Hence higher standards of insulation must happen.

If you do decide to better insulate your home you will discover there is a wide and potentially confusing range of insulation materials available. Table 3.3 using information produced by the Joint Research Centre (JRC) of the European Commission shows the thermal insulation market in Europe in 2014 by volume.[19] Over the past ten years the sale of nature-based insulation materials such as wood fibre has slowly increased from an extremely low level. This increase has varied from country to country with France now using 11 per cent and Germany 9 per cent nature-based insulation – in both cases wood fibre has the largest share. For the UK the 2014 figures have hardly changed.

Insulation material	Percentage of European market
Glass wool (also called glass fibre)	36
Expanded polystyrene (EPS)	27
Stone wool (also called mineral wool)	22
Polyurethane (PU/PIR)	8
Extruded polystyrene (XPS)	6
Nature-based materials, including cellulose, such as wood fibre, cork, hemp, wool, etc.	1

Table 3.3: The thermal insulation market in Europe in 2014, by volume
Source: JRC European Commission.[20]

However, while it *seemingly* does not matter what material you insulate your building with a closer look at the pros and cons of the various insulation materials available reveals some interesting disparities, which – in the battle against climate change – are worth considering and acting on. They are worth considering for the same reason you would recycle an aluminium drink can. Recycling a single can will not save the planet, but if *most* drink cans are recycled, it will help. Similarly, your decision to choose insulation material X over insulation material Y will not on its own halt climate change, but if millions of homes all over Europe were to choose X over Y – and if X *does* have a higher climate benefit over Y – then the multiplying effect will make a quantifiable difference. If you can modestly insulate your home, you may save 500kg CO_2 per year. If 10 million UK homes did the same (the UK has 24 million residential buildings) it would save around 1 per cent of the country's territorial carbon emissions. Working out which insulation materials are most effective at reducing carbon emissions and, as with building materials more generally, their level of embodied carbon, is of major importance.

While the climate consideration is crucial the choice of insulation will also likely involve practical considerations such as weight and volume. For example, if your loft space is difficult to access, typically through a small entrance in the ceiling above a stairwell, then, for safety reasons, you may well opt for the lightest insulation or one that can be compressed and still reform. If you live in a house made from solid stone with small rooms you may well not wish to use a relatively thick interior wall insulation as this will make your rooms noticeably smaller. If you use a loose-fill wood fibre insulation made from recycled paper (one of several cellulose materials) it will have more limited applications (e.g. timber frame walls behind the breather membrane or lofts). (It is worth noting that badly installed insulation can be a major problem in both old and new builds, causing more problems than it solves, for example, damp).

The manufacture of almost all insulation materials, irrespective of whether they are nature-based materials or not, involves large amounts of energy. Most European energy is still sourced from fossil fuels so potentially most insulation materials will have high levels of embodied carbon (derived from the burning of fossil fuels to create the energy to make the insulation – as we saw with concrete in Chapter 1.) Consequently, we find ourselves aspiring to reduce fossil fuel use by better insulating our homes, but in the process we end up burning more fossil fuel to make the materials that will reduce our fossil fuel use! This conundrum is increasingly widely recognized, as here by Mantle Developments:

> In Canada, insulation is an essential building material. Insulation materials reduce operational energy and emissions by limiting energy flow through a building's envelope. But manufacturing insulation can emit significant . . . carbon. As energy codes call for increased efficiency, one of the main solutions is to add more insulation. However, the embodied carbon impact of that additional insulation is often overlooked.[21]

Although on the face of it this undermines our attempts to reach net zero housing emissions, it *definitely does not negate insulating our buildings*. Even those insulation materials that have the highest level of embodied carbon will over a relatively short time span (2–3 years) start delivering a considerable saving on the overall amount of fossil fuel that is burnt to heat the building over its lifetime (75+ years). This is the "carbon payback period", defined by the length of time it takes for the avoided emissions resulting from the energy savings delivered by the insulation to be greater than the embodied carbon in the insulation resulting from its manufacture.

Given the climate crisis we must seek to use the insulation materials that most successfully reduce the operational carbon emitted *and* have the lowest level of embodied carbon. Simply saying "Just use any insulation as they all save on CO_2" will not help to *maximize* the climate benefits of insulation. Why? The answer is in the maths. A small amount is just that, a small amount. But if millions of homes reduce their CO_2 emissions by a small amount then collectively we will have saved a large amount of CO_2.

If we estimate the lifespan of a typical house built today to be 75 years, if the carbon payback period between purchasing two different insulation materials is "only" six months it would be understandable if the individual homeowner discounted this in their decision-making as it equates to less than one per cent of the house's lifespan, so from their perspective it is not a very big carbon difference. But if you look at it from the perspective of the house builder who may be building 4,000 such homes in a single year, you then find yourself looking at a much bigger figure: six months × 4,000 homes = 2,000 years of additional potentially unnecessary CO_2 emissions. So, while the individual houseowner's purchasing choice may seem insignificant, the house builder's choice is far more significant – it is back to the recycled drink can analogy. There is no financial benefit to the drinker, but a huge multi-million pound recycled aluminium business – and the carbon emission savings that go with it – is dependent on the single drink can being recycled.

In determining which insulation materials have the lowest embodied carbon we need to consult the Environmental Product Declaration (EPD) of the insulation under consideration. An EPD is a standardized document providing information about a product's environmental and human health impact.[22] Here we find ourselves in a difficult area. EPDs are the best we currently have and we do not want to disparage them. Far from it, rather we want to improve them. However, EPDs can report quite wide differences of assessment on two insulation products that appear to be very similar in design and material composition. This results in each type of insulation having a spectrum of EPD results across the different market brands producing this type of insulation. For instance, the same insulation material can be formulated with different densities, conductivities and with different facing materials, all of which will alter thermal performance.

Similarly, one type of stone (mineral) wool insulation produced by Company X could have a "high" level of embodied carbon and another apparently very similar stone (mineral) wool produced by Company Y could have a "medium" level of embodied carbon. Although this is confusing, *it does not mean the EPDs are wrong*, rather it suggests the *same* material used to make a similar product produced in different countries or factories can have different levels of embodied carbon due to different energy sources being used. This can differ between country energy grid mixes, or even factory suppliers in the case of electricity, or in terms of the fuel sources used for the heating process.

What if the materials used to make the insulation are *different*? If we compare the stone (mineral) wool insulation produced by Company X with a wood fibre insulation produced by Company A we might well *expect* to see a difference in the amount of embodied carbon between the two products as reported by the two EPDs. We could reasonably expect this difference because they are made from different materials.

But at this point we need to be careful. An understandable presumption to make is that the nature-based wood fibre insulation would have a lower embodied carbon than the stone (mineral) wool insulation. After all, the stone (mineral) wool has been made by heating rock to 1,300–1,450°C to melt it in a process that will have involved the burning of fossil fuels. The molten rock is spun in a machine – a bit like candy floss – and then it is cooled to produce blanket-like sheets to lay in your loft. But surprisingly it *may not* have a higher level of embodied carbon. In the same way, the wood fibre insulation may also have been made using a large amount of energy sourced from the burning of fossil fuels. For example, the manufacturing process could have been what is termed a "wet" process, where at the end the wood fibre needs to be dried. This drying, while not to the same very high temperature as needed to manufacture the stone (mineral) wool, could be a lengthy process. Hence manufacturing stone (mineral) wool at a high temperature over a short process compared to manufacturing wood fibre at a low temperature over a long process could see either product having the higher level of embodied carbon – or about the same.

There is however a sustainability issue at play here too: you cannot grow more stone. Stone while a natural material is not renewable, hence we have a finite amount of it particularly close to our cities and manufacturing centres and it is heavy to transport. Timber conversely is a renewable resource, is sustainable and it is lighter to transport.

Polystyrene (expanded and extruded) and polyurethane are both a type of plastic and are therefore oil based. The resulting insulation materials are highly efficient at reducing energy use and they have the benefit of also being light, but their embodied carbon levels are just as high as their rival materials and noticeably higher in the case of polyurethane, as we shall see.

In a significant piece of research conducted in 2017 Hill, Norton and Dibdiakova examined 60 published EPDs on insulation products.[23] Their resulting paper served in the first instance to illustrate the complexity of this area and why consumers are so often baffled by the details of product information. A customer requesting loft insulation with a low carbon footprint could be offered mineral wool in one DIY store, wood fibre in a second and both in a third. That each specific insulation product sits somewhere on a spectrum of possible EPD scores for its embodied carbon explains this situation. As this spectrum is quite broad it becomes difficult to generalize as to which type of insulation material will score best or worst from an embodied carbon perspective.

The insulation product identified in the research that had the lowest embodied carbon was a cellulose insulation. Cellulose is a complex carbohydrate derived from the cell walls of plants and is commonly obtained as a white stringy substance from vegetable matter, including wood, which is used in making various products such as paper. In this case the cellulose concerned was recycled paper. Recycled paper makes for very good insulation. As paper is made from wood fibre, insulation made from recycled paper is technically a wood fibre insulation. However, it is significantly different from other types of wood fibre insulation that involve "using fibres obtained from wood chips by chemical or mechanical separation technique".[24] For this reason, in the construction industry, wood fibre insulation and recycled paper insulation are seen as two separate insulations although in Hill, Norton and Dibdiakova's research they are grouped together under the umbrella term "cellulose".

Insulation made from recycled paper also has the distinct advantage that its embodied carbon has already been counted when it was first manufactured as paper. Hence using a recycled paper insulation has the big advantage of having an agreed low level of embodied carbon. It does still have some because it will have been processed (involving some energy), transported to the factory and then the customer.

Others have also sought to rank the embodied carbon levels of different types of insulation. In Canada, KPMB Architects produced a 2021 report which noted: "Like many architects, we have begun to pay much closer attention to the embodied carbon associated with the materials we are specifying. All other things being equal, selecting a material with a lower global warming potential (GWP) is preferable. However, at this early stage, not many of us have a strong intuitive sense of how meaningful various GWP values might be".[25]

Research by Jamie Bull has examined the embodied carbon of different types of insulation material.[26] Each material was asked to perform the same insulation task and to have the same surface area. It was first noted that this led to a difference in the degree of thickness of the material needed. What also varied was the amount of embodied carbon that each comparably performing material contained (see Table 3.4).

In the Bull calculations (Table 3.4) polyurethane is clearly in a league of its own with a noticeably higher level of embodied carbon. It should also be noted that "woodwool board" is not made from wood and sheep's wool but from shredded timber bound together with a cement paste.

As already highlighted a key consideration is the source of the energy used to make the insulation? Is an insulation product with a high level of embodied energy/carbon derived from wind energy a "greener" product than one produced with a medium level of embodied energy/carbon derived from burning a fossil fuel? This information needs to form part of the climate consideration process as to which insulation product to opt for. In a similar vein it could well be that a cellulose insulation made from wood fibre does indeed have a high level of embodied carbon but that this energy was derived from burning those parts of the timber log that were

Insulation material	Embodied carbon per m² of wall
Polyurethane unfaced	185.94
Polyurethane expanded	160.06
Rockwool (200kg/m³)	44.68
Woodwool board	39.56
Mineral wool	39.95
Foam glass (140kg/m³)	34.31
Foam glass (130kg/m³)	34.31
Cellular sheet glass	31.67
Rockwool (100kg/m³)	21.68
Cork slab (300kg/m³)	18.39
Expanded polystyrene	17.37
Rockwool (60kg/m³)	13.01
Flax	12.64
Cork slab (160kg/m³)	7.76
Cork (board)	7.13
Cork (general)	4.9

Table 3.4: Embodied carbon ($kgCO_2e$) per m^2 of wall

Source: Jamie Bull.

Note: The m^2 in the column on the right is referring to the functional unit (a square metre of wall with the required U-value) while the m^3 is referring to the density of the specific product.

not needed to make the wood fibre insulation. I hesitate to call it "waste wood" because there is no such thing since every part of the tree can be used, often for multiple purposes. Nonetheless many would see wood fibre insulation produced using the green energy generated from burning "waste" wood preferable to insulation made from burning fossil fuels.

In future we are likely to see manufacturers of all types of insulation switching to green energy, – solar, wind or biomass – as a way of lowering their embodied carbon values.[27] From a climate perspective this would be a good thing. An early mover in this area is Flumroc a Swiss manufacturer of mineral wool who from 2024 will source their energy from hydropower.[28]

That said we need to be mindful that green energy used to make a product is not then available for another use. An indiscriminate use of energy simply because it is green is not helping the climate. Rather green energy needs to be used as sparingly as possible hence the manufacturers of nature-based insulation materials such as wood fibre need always to be working to reduce their energy consumption even if it is from a green source. This approach also applies to engineered timber products (see Chapter 4) where EPDs are signalling a relatively high level of energy use, albeit often green energy. The efficient use of green energy in the battle against climate change is an imperative as it will increase the speed at which we can decarbonize the wider energy grid and drive down fossil fuel use. And what of nuclear power? Is this not also green power and therefore a better alternative to using fossil fuels? In France there are mineral wool factories where the energy source is 80 per cent nuclear electricity.

Insulation, as we have seen in this chapter, is a potentially confusing subject but if we are serious about tackling climate change and reducing the carbon footprint of our built environment then we must choose the insulation products that have the lowest level of embodied carbon while also factoring in the source of energy used to produce it. EPDs show the differences in the manufacturing process being used, and while they are not perfect, there has been a huge effort to harmonize them including stricter rules on what can and cannot be included making them far more useful for comparisons. But it is also the performance of the materials that is important.

Wood fibre may not be any better if you only consider the thermal insulating effect but most are semi-structural and when used on an interior wall they are solid enough to hang pictures on or support roofing slates in external roof situations. Additionally, as they are often denser, they have great noise insulating properties and they provide hygroscopic buffering, i.e. the ability to regulate relative humidity in homes and buildings by absorbing and desorbing moisture – a property that is often overlooked but can make a big difference to how much heating is used in a house as a damper home will feel colder.

But is this deliberating over which type of insulation to use from a climate perspective as complicated as it seems? No, it is not, because we have another very significant consideration to make when choosing an insulation material: does it *store* carbon?

Here the evidence is clearcut (see Table 3.5). None of the mineral (stone), glass, polystyrene or polyurethane insulation materials store carbon. You could look at a thousand EPDs for these products and none of them will reference any stored carbon. Conversely all the nature-based insulation materials, including wood fibre, recycled newspaper, hemp, cork, straw, sheep's wool, flax, cotton, etc. irrespective of their type or method of manufacture will be storing carbon. Woodwool board, as noted earlier, is a hybrid wood-cement material and as such exhibits aspects of both materials, i.e. it can sequester carbon in the wood but also has a relatively high degree of embodied carbon mainly in the cement. If Bull's considerations had included a 100 per cent wood-fibre insulation we can presume it would have scored second only to cork on sequestered carbon and would have had a lower level of embodied carbon than the woodwool board.

Currently we do not have a Europe-wide agreed method of blending the embodied carbon value of an individual insulation material, let us call it an "apple" measurement (in the right-hand column in Table 3.5), with it is ability to store carbon value, which we will call a "pear" measurement (in the left-hand column). The reason we cannot generate a single value is because apples and pears are different. The customer will have to make their own judgement as to which to prioritize and to what degree. It may be that in future when stored carbon is itself monetized the "pear" will be a more influential factor than it is now. This will certainly be the case if this monetization of stored carbon results in nature-based insulation materials (including cellulose products such as recycled paper, cork, wood fibre) becoming cheaper than the alternatives.

Sequesterd carbon per m^2 of wall	**Insulation material**	**Embodied carbon per m^2 of wall**
Higher is better		*Lower is better*
177.47	Cork slab (300kg/m^3)	18.39
68.84	Cork (board)	7.13
47.32	Cork (general)	4.9
74	Cork slab (160kg/m^3)	7.76
74	Woodwool board	39.56
13.63	Flax	12.64
0	Glass fibre quilt	3.8
0	Rockwool (60kg/m^3)	13.01
0	Expanded polystyrene	17.37
0	Rockwool (100kg/m^3)	21.68
0	Cellular sheet glass	31.67
0	Foam glass (140kg/m^3)	34.31
0	Foam glass (130kg/m^3)	34.31
0	Mineral wool	39.95
0	Rockwool (200kg/m^3)	44.68

0	Polyurethane expanded	160.06
0	Polyurethane unfaced	185.94

Table 3.5 Stored/sequestered and embodied carbon ($kgCO_2e$) per m^2 of wall

Source: Jamie Bull.

This would be game changing. In the opinion of the Vice-Chair of the UK's Alliance of Sustainable Building Products, Mark Lynn, the crux is whether we place a benefit and positive impact value on sequestered carbon.[29] Given that insulation and many other construction products have a designed life of many decades, that alone should justify some sort of value for the sequestered carbon because mitigation is needed today, and carbon locked up today is more beneficial than carbon locked up tomorrow. Furthermore, as we move to ever greater circularity in the design, manufacture and use of products then the probability of sequestered carbon being re-emitted as CO_2 at the end of life becomes ever more remote (see Chapter 7). Used in the correct setting, natural materials have the capacity to remain useful for many centuries not just decades. It is reasonable to argue that you should not give a value to the entirety of the carbon in nature-based materials but it is absurd to think that it should have no value whatsoever, which is the position of several of the green NGOs.[30] Even a conservative estimate of value for sequestered carbon tips the balance in favour of nature-based materials in many end-uses including insulation.

We have three urgent points of action if we are to meet our net zero targets by 2050: (1) We need to increase dramatically the amount of renovated/retrofitted homes being insulated. *A high standard of insulation in our entire built environment is the single most effective way of driving down energy consumption and fossil fuel usage.* (2) At the same time, we need to avoid using those insulations with the highest levels of embodied carbon. And (3) We need to prioritize using those nature-based insulation products that have the ability to store the most carbon, such as wood fibre insulation.[31]

As the European Panel Federation, the European trade association representing wood-based panels, put it at the EU Bioeconomy Conference in October 2022:

> With energy prices rocketing the number one recommendation is "insulate your loft". Yet, only 1% of insulation materials are currently nature-based, the other 99% are all fossil fuel based. This needs to change and, as European Commission Vice-President Frans Timmermans highlighted when addressing the Bioeconomy Conference earlier today, wood-fibre is the sustainable material that can do this and additionally it will safely store carbon.[32]

The word "safely" may raise concerns as to whether using wood fibre is a fire risk. Chapter 5 will address in detail fire risk and wooden buildings. To summarize, it is wood's ability to store carbon in the built environment that makes it central to our efforts to combat climate change. The average three-bedroom house, insulated by wood fibre to a high standard, can easily store two, three, four or even *five* tonnes of carbon. Multiply this for millions of homes across Europe and beyond and you have the opportunity to store in the built environment phenomenal amounts of carbon previously sequestered by trees in the forest. We know that the forests of the EU currently store and offset the equivalent of 10 per cent of the continent's carbon emissions. Our target should be to hit a point as soon as possible where Europe's *building stock* stores as much carbon as the forests themselves. The quickest way to achieve this is via the extensive and widespread use of wood fibre insulation. Combine this with a significant increase in the number of both timber-framed homes and tall/large engineered-timber buildings, the subject of Chapter 4, and we will be well on track for a net-zero built environment. As we've seen, this approach lies at the heart of Schellnhuber's and his colleagues' thinking: by using wood-based materials in the built environment, in ways which maximize the amount of carbon stored and generate low emissions, we can grow the human-made global carbon pool while at the same time reducing the CO_2 emissions generated by construction.[33]

Huge amounts of time, energy and money are going into researching and trialling carbon capture and storage via industrial processes such as carbon storage underground in former oil or gas fields. And yet right in front of our noses is this opportunity to store carbon in the loft and in the walls of the *very homes we live in*, through wood-based insulation. We are obsessed with finding a *technical* solution to a problem when nature has already handed us the answer.

A hopeful sign in March 2023 was the decision by the Portuguese-Chilean wood-based materials group Sonae Arauco, one of the main producers of wood-based panels in the world, to invest €100 million in the construction of a new production line for wood fibre insulating materials at their Meppen site in Emsland, Germany.[34] Their wood-based products are already storing around 2 million tonnes of CO_2 globally.

Another hopeful sign is what is happening in Wales where they are working to set up the UK's first wood fibre insulation factory using local timber. As announced by the Welsh Assembly's Minister for Climate Change Julie James:

> The manufacturing of woodfibre insulation not only provides an added-value market for our smaller diameter logs and residues, but also provides a key product for the decarbonization of our homes. In particular, woodfibre insulation will support development of our vibrant timber frame manufacturing sector and is particularly important for the low carbon retrofit of our existing homes, many of which are solid wall. We are confident that locating a woodfibre insulation factory in Wales will complement our programme and provide a lasting legacy for the construction, manufacturing and forest industries sector.[35]

The only snag with Mother Nature's solution is that the built environment is huge and expanding all the time. This in turn means that if we maximize the use of wood, in both new-build and in renovation/retrofitting insulation, we are going to need a colossal amount of it. Chapter 7 will examine whether we could have enough sustainable wood to do the job.

4

Wooden skyscrapers

Fact: *The world's tallest wooden building is 18 storeys high, built with engineered timber which has a structural strength equivalent to that of steel and concrete.*

As an 11-year-old growing up on Tyneside in the 1970s, my mecca was the model kit shop for hobbyists on Nile Street, North Shields. It was a round trip of four miles on foot from our house but it was a regular Saturday morning expedition. Run by a mother and son, the shop was chock full of model railways, plastic kits, radio-controlled planes, glues, paints and balsa wood – from floor to ceiling. Buying an Airfix kit was a must. The company's vast array of model planes was the subject of much discussion, catalogue browsing and indecision. It is the British de Havilland Mosquito that for some reason lingers in my memory (Figure 4.1). Maybe it was a harbinger of what was to come because the Mosquito was famous for being made from timber, known affectionately as the "Wooden Wonder". It was pretty obvious that the flimsy aircraft of the First World War were made from wood but surely the more robust, bigger and faster aircraft of the Second World War were made from "tougher" materials? No. Wood remained a key structural component of many aircraft including, in addition to the Mosquito, the Russian Ilyushin il2 and the German Focke-Wulf TA154. In the 1940s it was acknowledged that wood was a strong enough material to build aircraft that could fly at 400 miles per hour and be subject to huge g-forces. A signal perhaps of wood's potential as a structural construction material?

Figure 4.1 Airfix Mosquito fighter bomber model kit
Source: Airfix – Hornby Hobbies Ltd 2023.

Of course, wood has been used to build homes for thousands of years. But the majority of these buildings have always been relatively small, as in not particularly tall – the one- and two-storey homes of rural and suburban North America: think "Little House on the Prairie", the family home of "the Waltons", through to today's aspirational clapboard homes of Cape Cod or The Hamptons. There were exceptions. For instance, and still standing today, the 67-metre-tall Yingxian pagoda in China built in AD 1056 is a wooden structure. The Norwegians have the 29-metre tall Heddal Stave Church built in the early thirteenth century. In the nineteenth century in both North America and Europe large warehouses were being built next to ports, which while stone clad on the outside, actually concealed an inner timber structure of large wooden posts and beams. Such buildings can still be found today in cities like Vancouver.

It was however the arrival of the Industrial Revolution, which saw coal replacing wood as a fuel and the advent of the mass production of iron and steel, that resulted in timber's declining use in big building construction. This decline was exacerbated by perceptions of wood's low durability and combustibility following major city fires worldwide including London, Paris, Chicago, San Francisco and New York.[1] But if we look more closely at the Mosquito's construction, we see an early deployment of a technique which has, over the past 50 years, revolutionized wood's potential in the construction of buildings in our cities.

The technique is not particularly difficult. It is beautifully simple. You probably know it as plywood. This, however, is the "grown up" version of plywood known as "engineered timber" in Europe or "mass timber" in North America. The main product comes as a panel and is known as cross laminated timber (CLT) (see Figure 4.2). It consists of not less than three cross-bonded layers of timber (lamellas) typically ranging in thickness of 2–4.5 cm, although they can be as thick as 8 cm. The timber has to pass a strength test and is then glued together in a press which applies pressure over the entire surface area of the panel. CLT panels have an odd number of layers (three, five, seven, nine) which may be of differing thicknesses; layers are arranged symmetrically around the middle layer with adjacent layers having their grain direction at right angles to one another. The overall thickness, as well as the loadbearing performance of the composite panel, is determined by the build-up of the individual laminates. CLT panel thicknesses are typically in the range of 8–20 cm but can be as thick as 35 cm.

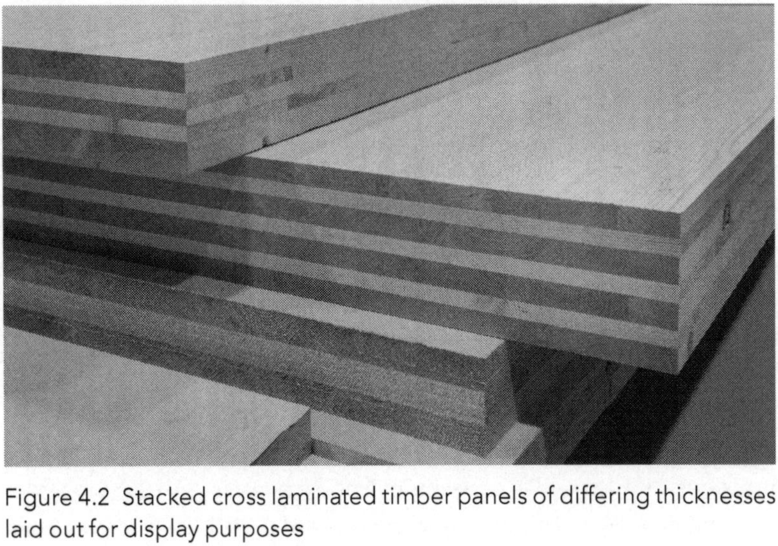

Figure 4.2 Stacked cross laminated timber panels of differing thicknesses laid out for display purposes

Source: Stora Enso.

The most commonly used wood species for CLT in Europe is spruce. In North America they have a more expansive approach that includes three main categories of timber: spruce-pine-fir; Douglas fir-larch; and southern pine. CLT panels can become walls, floors or ceilings. CLT's game-changing nature is that it has the structural strength of steel and concrete combined with a higher strength to weight ratio i.e. given the same structural task to perform your piece of CLT will be lighter than your piece of steel or concrete. This means we can use CLT to build far taller timber buildings than the traditional two-storey home inhabited by the Waltons.

Figure 4.3 Laminated veneered lumber samples

Source: Metsä Group.

CLT has two siblings: laminated veneered lumber (LVL) (Figure 4.3) and glued laminated timber (glulam) (Figure 4.4). Both are constructed in a similar way to CLT – i.e. layers of wood – however LVL is made from much thinner layers of wood which have been "peeled" from a tree trunk. As the layers are thinner, between 2–3 mm, there are more of them, say 17.

In turn it is a lighter material by volume than CLT and can perform a similar structural role. Initially it was not used to build beyond five storeys, but this is changing and it is now being used to build to greater heights especially when combined with CLT.[2] LVL is the strongest of the engineered timbers. It is twice as strong as steel in proportion to weight and 2.9 times as strong as concrete in compression, i.e. able to carry a load, usually above. What it lacks in beauty compared to CLT is made up by its material efficiency. In other words, you can build a five-storey LVL building with less wood than the same-sized CLT building.

Figure 4.4 The glulam roof at the Macallan distillery in Scotland
Source: Wiehag.

Glulam is the most versatile of the engineered wood products. It can be shaped into forms ranging from straight beams to complex curved ones, often used in roof structures. A mechanized, hydraulic press is used to curve the layers of wood with the curve preserved once the glue has dried.

The mention of glue has hopefully rung a few alarm bells and rightly so. We have managed to invent some horribly noxious glues over the years, including solvent-based adhesives that are harmful to the environment generally and humans specifically. In response much work has been undertaken to improve the glues used to make engineered timbers and to ensure they are neither emitters of nasty chemicals or more flammable than the wood they hold together.[3]

Engineered timber is not new. Glulam has been available in Sweden for over 100 years. The beautifully curved glulam roof arches at Malmö Central Station were created in 1924. CLT is the newest of the engineered timbers and was developed in Austria in the 1990s and then commercialized in the 2000s. The first modern engineered timber tower was Murray Grove in Hackney, London built in 2009 (see Figure 4.5) and was the brainchild of architects Andrew Waugh and Anthony Thistleton. It was the world's first tall urban housing project to be constructed entirely from prefabricated engineered timber in the form of CLT, from the load-bearing walls and floor slabs to the stair and lift cores. It was a deserved winner of the Royal Institute of British Architects 2009 President's Award for Research.

Figure 4.5 Murray Grove, Hackney, London
Source: Waugh Thistleton Architects (Will Pryce).

Completed within 49 weeks, and delivering 29 fully insulated and soundproof apartments, the project demonstrated for the first time that CLT had the potential to be a financially viable, environmentally sustainable and beautiful replacement for concrete and steel in high-density housing. Today many of the first residents are still living there and the high levels of insulation mean energy bills are low, much appreciated as the price of oil and gas rise.

Following Murray Grove, the Treet or Tree House in Bergen, Norway, built by the BOB housing association, became the next tall wooden building to register among architects. Built on the harbour side it rises to 14 storeys. At 49 metres it is almost the height of Nelson's Column, overtaking Murray Grove as the tallest wooden building in the world when completed in 2017 (see Figure 4.6). I was able to go and see it when it was half built.

It was a journey in two parts. I first took a short trip to the north bank of the River Tyne in my hometown of Newcastle and gazed across the water to view Dunston Staithes. The Staithes is a wharf, built from large pieces of ordinary tree timber, and opened in 1893 to enable trains carrying coal from the Durham coalfield to be unloaded onto ships that would take it to London and further afield. For many years, and maybe until very recently, it was reputed to be the largest wooden structure in Europe – or, to be more accurate, having the most wood in a single structure. It seemed a fitting place to visit before crossing the North Sea to witness the construction of the Tree House.

When I arrived at the Tree House the first thing I noticed was the outer structure, the frame in effect, which was clearly made from CLT. And then rising through the centre were the lift shaft and staircase, again all CLT. I watched them lowering prefabricated wooden units. These modules, built in Estonia by a company called Kodumaja, had been delivered by ship to the site where they were craned off the vessel straight into the inside of the frame. These units already contained windows, kitchens, bathrooms, pipes, electrics and so on, and they were simply clipped into place before the pipes and wires were connected. With everything factory-made, prefabricated and then delivered to site for assembly, the tolerance levels are fantastically high meaning everything does fit together – no gaps!

Figure 4.6 Tree House, Bergen, Norway

Source: BOB.

Cards on the table – even major wooden structures need concrete foundations. Concrete, despite being highly carbon intensive, is exceptionally good at certain roles. When it comes to foundations for tall wooden structures there really is no alternative material at present. Small wooden structures can be built on wooden piles driven into the ground, or on stone foundations, but if you want to build big in wood, on the scale of the Tree House, concrete is the appropriate material, not least because it remains immune to water damage for decades. However, as wood is a lighter material to build with than both concrete and steel, the foundations do not need to be as extensive. Hence tall wooden buildings typically use less concrete for their foundations. In a similar way we still need steel since most of the connections that are needed to join the sections of engineered timber together are made from steel. So yes, the Tree House has concrete foundations but from ground level upwards structurally it is all wood – even the lift shaft, just like Murray Grove.

All tall buildings – irrespective of what they are made from – will sway when the wind gets up. Structurally this is not a problem, in fact they are designed to sway. Like a tree on a windy day swaying is the building's natural way of coping. However, tall wooden buildings because they are lighter can sway more and residents at the top of a tall wooden building can find this swaying disconcerting. Consequently, anchoring the building with a concrete lift and stair shaft running through the heart of the building can make sense. Alternatively, a single concrete floor half-way up the building can reduce the sway. But the argument that lift shafts need to be made from concrete to get the tolerance levels right enough to be able to fit the lift mechanism is a misconception. A wooden lift shaft built in factory conditions (dry and indoors) is more likely to be built to a finer degree of tolerance than one made of concrete which has been poured on site, floor after floor, in all weathers.

The Tree House did not need a concrete lift shaft, or a concrete floor at the mid-point, to provide the weight to prevent swaying. This was because it was in a valley at the lowest point which afforded some wind protection, further shelter was provided by a large road bridge, shielding it from the prevailing wind and at 14 storeys it was slightly short of the point at which swaying in high winds becomes an issue.

Two things in particular struck me about my guided tour of the lower half of the Tree House while it was under construction. On the sixth floor I was taken into one of the completed flats which, due to the design, had a structural CLT beam, a good 60 × 50 cm, running at a 45° angle through the living room (see Figure 4.7). "Does this mean that this flat will be a little cheaper than the others?" I naively asked. "No", came the reply. "If anything, it's quite the opposite as people like the exposed timber both visually and to touch". That's exactly what I had seen and done myself on entering the flat.

The second thing that struck me was that each of the three-, two- and one-bedroomed flats that made up the building, as well as the studios where space was at an absolute premium, all had a sauna!

Despite the Tree House being some months away from completion, with a further seven floors to be constructed, the majority of the flats were for sale and already had offers on them. (There were also flats to rent – some at social rent rates – so it was clearly going to be a mixed community). Again, I made a miscalculation: "Presumably most of the deposits to buy have come from young people?" No, quite the opposite again. The majority of offers were from older people who had grown up in Bergen and, as they got older, had either moved to the suburbs or further afield to raise their children. Now retired they wanted a smaller place nearer to all the facilities, so they were moving back into town.

Figure 4.7 Inside the Tree House, Bergen, the engineered timber frame is clearly visible
Source: BOB.

Norway's Tree House held the record for being the world's tallest wooden building for all of two years before the completion of the 18-storey, 57.9 metre, Tallwood House at Brock Commons in Vancouver in 2017. The building, a student residence for the University of British Columbia (UBC), was designed by Acton Ostry Architects of Vancouver, with Hermann Kaufmann Architekten of Switzerland acting as tall wood advisors. The building's component parts were manufactured in factories away from the site and the process was an assembly rather than a traditional build, shooting up two floors per week. The project team described the assembling of the building as "like Lego". The building uses a hybridized construction model with concrete lift shafts acting as stabilizing cores and prefabricated steel elements assisting the mass timber frame. Lateral and vertical stability is gained from the combination of these material elements. In order to meet Canada's new seismic (earthquake) design requirements steel connectors are fitted to the glulam columns to provide direct load transfer between the columns and a grid of CLT panels. Unlike the Tree House, Brook Commons Tower, at nearly 60 metres, has the potential to "sway too much in the wind for top floor comfort" and so a concrete floor was placed midway up the build to add weight to reduce swaying.

The Canadians held the record for the tallest timber building for three years before the Austrians took the accolade with their amusingly named HoHo building in Vienna at 84 metres and with 24 storeys (which I also managed to visit in November 2017 while it was under construction). But the Norwegians responded quickly winning the record back with Mjøstårnet[4] before the HoHo was even finished! While having fewer storeys than the HoHo building, "only 18", it is 60 cm taller. Although, if you look closely (see Figure 4.8) this extra height is simply down to some questionable wooden framing on the roof which you cannot help but think is only there to out-height the Austrians!

Norway will not lead for long as by 2026 residents should be able to move into the 100-metre tall load-bearing timber construction Rocket&-Tigerli, named after locomotives produced at the site, in the Swiss city of Winterthur, close to Zürich.

For some in the world of wood this ongoing, slightly Freudian, competition regarding height is a distraction from the serious business of building more with wood which would help tackle climate change. News stories about yet another even taller wooden building helps spread the word that building tall and big in wood is now possible due to engineered timbers, but a relatively small number of ever taller timber skyscrapers is not going to decarbonize the built environment; rather it will be thousands of mid-height, 10–15 storey, timber builds all over the world that will make the biggest impact because such buildings are by far the most common in the world's ever-growing urban environment.

Figure 4.8 Mjøstårnet, Norway, the world's tallest timber building in 2024
Source: Moelven.

Aside from the headlines about the tallest timber towers, the overall construction of large wooden buildings, especially mid-height ones, is booming around the world and the demand for CLT panels has increased dramatically. At this point a dose of pragmatism is called for. While a steady increase in the number of buildings that are made predominantly from timber, *substituting* for the concrete and steel that would otherwise have been used, would be a good thing from a climate perspective, it would also be a good thing if more concrete and steel buildings contained *some* timber, as opposed to none. Enter stage left, the hybrid building.

A good example of what is achievable can be found in Milwaukee, USA. Here the now completed Ascent Tower has six lower storeys of a reinforced concrete structure with a further 19 storeys added on top made from CLT manufactured by the Austrian company Wiehag (see Figure 4.9).

Hybrid thinking is important as it can help us work out how to use each material in the most climate efficient way. As noted in *Tomorrow's Timber*, "a lot of research is going into the development of hybrid building systems in which timber is complemented by concrete or steel for maximum synergy between structural integrity and efficient material use, while maximizing the timber percentage to reduce the carbon footprint".[5] In some buildings this could mean steel girders supporting CLT floors that themselves have a concrete topping layer. This layer can help protect the CLT from being damaged by rainwater while being transported or during construction until the building is watertight. It can also add the weight needed to prevent taller engineered timber buildings from swaying too much in high winds (due, as noted earlier, to the lightness of the wood). Finally, such a layer can aid sound-proofing.

What we are really aspiring to here is achieving a position of parity for wood in a conservative construction world that for 150 years has been dominated by concrete and steel. Not until wood joins concrete and steel as a core part of the structural palette for architects, engineers and construction firms will we see a noticeable reduction in the climate footprint of the built environment.

Figure 4.9 Ascent Tower, Milwaukee, USA, a hybrid concrete and engineered timber build

Source: Wiehag.

An interesting variation of the hybrid build outlined above, concrete lower storeys with timber upper storeys, is an approach known as "build on top". Here an existing concrete and steel, or brick and block, building can have additional storeys of wood built on top – an extension. This approach can be most useful in urban areas where space is at a premium and land for new builds is scarce and expensive. As wood is a relatively light building material it may well be possible to add three or four additional storeys in wood as opposed to only one or two storeys if more weighty carbon-intensive materials such as brick and block – concrete and steel – were used.

A fine visual example of the build-on-top approach can be found in Washington, DC where a seven-storey existing build had an additional three storeys of mass timber floor space added on top (see Figure 4.10). One of the reasons why this is often possible as an approach is because existing reinforced concrete-based buildings have often been over-engineered, i.e. they could actually have been built taller in the first place due to their substantial foundations and the structural strength of the build.

Anecdotally, the concrete and steel office blocks of the 1960s and 1970s were all designed to carry the weight of hundreds of filing cabinets full of paper. As anyone who has ever tried to move a full four-drawer metal filing cabinet can tell you, they are very heavy! Today, with this burden lifted and the information instead stored in the cloud rather than in the cabinet, many of our existing buildings now have the strength to accommodate much needed extra capacity on top. For example, in Stockholm, located in the old industrial neighbourhood, the 1920s textile factory Trikåfabriken has been transformed into a modern and sustainable office complex. It now has a five-storey wooden framed extension on top of the old factory. The extension's architects, Tengbom, calculate that CO_2 emissions were kept 85 per cent lower than if they had completely demolished the old building and built a new one in concrete and steel.[6]

An underappreciated outcome of using more wood in construction is that it also improves air quality. Concrete is delivered to site in that

Figure 4.10 An engineered timber "on top" extension to an existing reinforced concrete framed building in Washington, DC
Source: Maurice Harrington, Sisson Studios.

ubiquitous urban vehicle, the "cement mixer". But, because it is a very heavy material, you need a lot of deliveries. Wood is notably lighter, hence far fewer deliveries are needed, i.e. to complete the same construction task – installing the fourth floor – the necessary wood can arrive on one lorry or the necessary concrete can arrive on four lorries. This is the principal reason why the recent construction of the Black & White Building in Shoreditch, London, designed by Waugh Thistleton Architects – using engineered timber – involved 80 per cent fewer truck deliveries than an equivalent concrete building. In 2020 nine-year-old Ella Adoo-Kissi-Debrah from south-east London became the first person in the UK to have "air pollution" listed as a cause of death by a coroner.[7] All heavy goods vehicles, whether they carry concrete or wood, pump out exhaust fumes that are a cause of air pollution. Fewer lorry journeys mean less air pollution caused by burning diesel – a major cause of climate change.

Lorry tyres are also a problem. As reported by the Chartered Institute for Environmental Health: "Tyres are made from synthetic rubber, a derivative of crude oil, and contain a number of toxic organic compounds, including known carcinogens. As they make contact with the road, they release tiny particles measuring less than 23 nanometres, which pollute the air, soil and water".[8] By choosing wood we are reducing the number of journeys made by heavy goods vehicles. This, in turn not only reduces CO_2 emissions, it also reduces air pollution in general, which is currently believed to be responsible for between 28,000 and 36,000 early deaths each year in the UK.[9]

But the Black & White Building's most impressive attribute is that its construction used 40 per cent less carbon than a comparable structure. Why? Because it's made of wood.[10] The structural frame is made from beech LVL from Germany, being used in London for the first time at scale and the walls and floor slabs are CLT from Austria.

To conclude: our inner cities, where space is at a premium and the average height of buildings is taller, no longer have to be the sole building preserve of concrete and steel. Engineered timber enables us to build big and tall with wood in a way that was unheard of just a few years ago. Talk about wooden buildings need not conjure up images of log cabins. The age of the wooden skyscraper has arrived – even if most people are yet to notice.

5

Fire: timber buildings and safety

Fact: *designed and built correctly tall timber buildings can provide a level of fire safety that is equal to or greater than that found in tall concrete and steel buildings.*[1]

Fire can kill. A badly designed and poorly constructed building, irrespective of the materials used, can be a death trap for those who live or work there. The 2017 Grenfell Tower disaster in London stands as a stark reminder of fire's ability to destroy lives and property. Seventy-two people died, primarily from toxic smoke inhalation, in the capital city of one of the world's richest countries with one of the best emergency services.

Tall buildings, including those built from timber, can be safe buildings with a low fire risk. The issue is not one of material, but of expertise. According to José Torero, a world-leading authority on the fire safety of large timber buildings, "Timber can be perfectly safe if it's done right. This is not a technological or technical problem. It is fundamentally a problem of competency. The problem is whether [building designers] are competent enough to be able to make a proper assessment, and that is a much, much more complicated question".[2]

In this chapter the three main types of structural timber build are discussed in the context of fire: post and beam, timber-framed, and engineered timber. The material that has been used to construct the frame of the building and the material that is preventing the building from falling down is timber and hence we use the term "structural timber".

Before discussing these different types of timber build in the context of fire, a word about saving lives and saving property. The first is infinitely more important than the second. Burnt out buildings can be rebuilt; burnt out bodies cannot. Hence when discussing fire in buildings the priority is the safety of the occupants and, if a fire does occur, the lives of the attending firefighters.

Post and beam

In my home city of Newcastle upon Tyne until recently the tallest wooden building was Bessie Surtees House.[3] Comprised of two merchants' houses, and built in the sixteenth and seventeenth centuries, they are five storeys high with a solid, heavy, timber frame of vertical posts and horizontal beams, referred to by architects as "post and beam". They are fine examples of Jacobean domestic architecture (see Figure 5.1). On the same street, a little further west, is The Cooperage built in 1430, also as a merchant's house and also with a post-and-beam construction. The ground floor ceiling timbers are reported to have come from a Dutch merchant ship that sank in the adjacent River Tyne. In recent years it became a brewery and then a pub.

Two hundred years ago across Europe and North America, the traditional way of building a timber-framed house was to fell a tree, remove the bark, square the timber off and then erect four or more perpendicular posts. At first floor height the posts would then be connected by horizontal cross beams. A second or third floor could then be added or more ground floor rooms necessitating more posts. This timber frame was structural, i.e. it kept the house up and stopped it from falling down. The timbers would be characterized by their large dimensions and their solidness – they were after all essentially tree trunks, not branches of trees. The frame would then be infilled with exterior and internal walls. The longevity of many of these buildings, as evidenced by Bessie Surtees House and The Cooperage, point to post and beam being an intrinsically sound form of construction.

Figure 5.1 Bessie Surtees House, Newcastle upon Tyne, an example of post-and-beam construction
Source: iStock; TraceyAPhotos.

Today houses are still built in this manner, but not in the case of the vast majority of homes with a timber frame. Rather it is a technique used to build bespoke homes in the higher price bracket. Oak is a popular choice of timber for this kind of build. Often the approach deployed is to build an extension to the existing older home, either to mirror the core structure's existing timber frame or to add an extra architectural dimension to, say, an old stone-built home. Either way per square metre of extension it is as expensive as it is beautiful and hence not particularly common.

I once shared a taxi in Newcastle with Councillor Ged Bell who for 30 years had been a firefighter with the Tyne and Wear Fire and Rescue Service. As we crossed the city we discussed building in wood. Bell remarked that he had always felt less worried entering a wooden building on fire than one made of steel or concrete because wood is more predictable in a fire than the other two materials. Concrete explodes at high temperatures. Steel melts and bends and the building can collapse. The taxi driver then joined in the conversation, telling us that he used to manage the fifteenth-century Cooperage pub on the Quayside where a few years earlier there had been a major fire. The day after the fire was extinguished, a fire inspector came to check the building. All the massive wooden beams were badly charred and it seemed to our taxi driver that the whole building would have to be demolished. However, as the fire officer explained to him, the outside of the wood may have been charred but underneath the wood's structural strength was still intact. All that was needed was for the charred surfaces to be scraped back until the unburnt wood was reached. And sure enough that was the case and The Cooperage is still standing today. Put plainly the large solid timbers used in post-and-beam structures are slow to catch fire and when they do, they will smoulder and char rather than blaze. Consequently in most cases the fire brigade will arrive in good time to extinguish the fire.

Timber framed

Today most buildings that use timber as their structural material, the material that stops them falling down, are the family homes we find throughout the suburbs and rural areas of North America, Scandinavia, and parts of the UK and Europe. These timber-framed homes are built using timber referred to as "sawn wood". The sections of sawn timber used are on average of smaller dimensions than what was used in their historic ancestor the post-and-beam structure.[4] A smaller timber section (smaller dimensions) is more vulnerable if exposed to fire than larger timber sections; consequently a modern European home built with a timber frame will have its frame shielded from fire by the use of a protective covering such as plaster board. This is why when entering a modern timber-framed house you do not see any exposed wood. It is virtually impossible to the untrained eye to tell if a modern family house has been built using a timber frame or not, either from the outside or the inside, as the frame is hidden both by an external covering such as brick and an internal covering skimmed with plaster.

The initial fuel for a fire in *any* type of building built from *any* type of material will be provided by the furnishings and contents of the building, which will burn before any structural elements ignite. Once the furnishings and contents are ablaze the temperatures reached are so high that no one can survive inside the room. This means that the structural material of a building has very limited influence, or even no influence, on safety of life. To be clear, the heat and the smoke from the burning furnishings and contents will kill you *before* the timber structure catches fire. For this reason, for example in Norway, sprinklers are required in all new residential buildings with three or more storeys, *irrespective* of the structural material.

The material(s) used to infill the timber frame, which provide the interior and exterior walls, including insulation, can increase the fuel available and this is significant regardless of the structural form, whether timber, concrete, steel, brick or block.

Timber frame is a well-established approach to house building. The timber frames can be assembled on site from sawn timber or in a factory (where they can also be infilled to form a ready-to-use wall panel) and delivered by lorry to site. Their familiarity extends to an understanding of how to ensure during construction that the wood is not exposed to any potential sources of combustion i.e. the risk of fire is mitigated. It can never be eliminated altogether as accidents can happen and arson can occasionally occur on a construction site, but building with timber frames is not seen by either fire authorities or insurers as any more problematic than house building with other materials. This remains the case when the houses are occupied.

On encountering a timber-framed house the most striking aspect, as already mentioned, is the lack of any noticeable wood. In most cases none of the timber frame will be exposed either inside or out. Externally it will be clad in brick, stone or render (see Figure 5.2). Internally the timber frame will be entirely encapsulated behind a dry lining of some type of plaster (UK) or gypsum (USA) board. Should a fire break out inside a timber-framed house the fire would first need to burn through the dry lining before the timber could potentially catch fire. The time taken for this to happen is such that fire fighters would typically arrive in time to extinguish the fire.

Figure 5.2 Akerman Road timber-framed social housing for Lambeth Council, London

Note: Modern timber-framed homes can be hard to spot as the timber is neither visible externally or internally.

If there was greater fire safety concern around timber-framed builds this would be apparent in the UK where England and Scotland have strikingly different approaches to house building. In England only 9 per cent of residential houses are timber framed whereas in Scotland the opposite situation exists – 92 per cent of houses are timber framed.[5] Why? According to the Structural Timber Association "Historically, structures would be built using materials that were immediately available – locally abundant and easily sourced. This notion is important as for Scotland, timber was a much more accessible resource than in England, where instead stone, sandstone and clay were the more easily obtainable materials".[6] Yet homes in Scotland are no more expensive to insure than their English counterparts, nor do the fire authorities express concern that homes in Scotland are more vulnerable to fire than in England. Similarly those house builders constructing homes with brick and block do not make claims about their homes being safer.

Compared to an engineered timber build, discussed next, the amount of wood in a timber-framed building is relatively small per square metre of the build compared to the other materials used. This results in what is referred to as a "low-fire load of timber" and when combined with a long history of safe use means that timber frame has been shown as a fire safe construction option for the low-rise housing and school building markets. Overall timber frame accounts for 25 per cent of the UK housing market and this share is growing.

Engineered timber

Engineered or mass timber is the latest arrival in the world of structural timber buildings. In Chapter 4 we examined how engineered timber (CLT, LVL, glulam) is enabling wooden buildings of much greater height and scale to be built. This greater height and scale gives rise to more demanding fire safety requirements than for smaller buildings because escape routes are longer and fire fighters may not be able to work from outside the building. However, these more demanding requirements have been successfully met in the newly built engineered-timber buildings, such as the 20-storey Sara Culture House in Sweden, which utilizes a combination of architecture, structural design, sprinklers and fire protection treatment.[7]

An essential technical difference between a timber-framed building and an engineered-timber building is that the latter has a much higher volume of timber within it per metre squared. And much of this timber may be exposed, unlike the timber-framed houses discussed above where the wood is typically encapsulated by some form of dry lining. This volume of wood is important. Not only are engineered-timber buildings much bigger than typical timber-framed houses, and so have more timber in them, they also use timber for more functions and so the ratio of timber to other construction materials is much higher. Most engineered timber buildings will have floors and roofs made from an engineered timber (CLT) as well as the structural frame (glulam or CLT). The walls are also often made from engineered timber (CLT or LVL) and can form part of the structural frame. Note that engineered timber is rarely used to build family homes, rather it is used to construct multiple occupancy blocks of flats, offices, hotels, schools, industrial buildings, etc. It can be used but it would be an unnecessarily expensive way of building a family home and would be an inefficient use of the timber, i.e. over-engineering.

From the perspective of tackling climate change this is a win-win. The increased amount of timber is *storing* more carbon than timber-framed buildings and is *substituting* for more of the highly carbon-embodied construction materials such as concrete floors, steel structural supports and brick and block walls (see Chapters 2–4). However, because engineered-timber buildings contain significantly more wood and much of it is exposed, when it comes to fire safety, they present a different set of issues than the millions of timber-framed houses that have dominated the housing construction market over the past 200 years.

The wood is exposed in engineered timber buildings for two primary and obvious reasons. First, people like the look of wood; it is aesthetically pleasing. They would prefer to look at a wooden wall rather than a plaster wall and they will pay to do so, hence the flats in the Tree House in Bergen that had exposed CLT beams were more expensive than those where the engineered timber structure was not visible (see Chapter 4). Second, encapsulating the engineered timber walls, floors and ceilings in another material (which serves no structural function) adds cost to the overall build. However, large surface areas, both horizontal and vertical, of exposed engineered timber do raise new and valid questions as to their fire safety.

Engineered timber, from a fire safety perspective, is very different to regular timber frame builds and we cannot simply lift across the thinking and approach to timber frame fire safety and apply it carte blanche to engineered timber. As already mentioned, the key difference from a fire safety perspective is that, compared to a timber-frame building, the "fire fuel is high" in an engineered-timber building simply because there is more wood and because this wood is often exposed. Large areas of exposed wood consequently present a challenge regarding the risk of fire spread. In turn the large amount of wood in an engineered timber build can provide a fuel source that lasts much longer than the room's contents which is not the situation in a timber-framed house. However, an engineered-timber structure can withstand fire much longer due to its thickness. After the initial stages of the fire, which will have resulted in the furnishings being consumed, the intensity of the fire in an engineered-timber structure will be dependent on the amount of oxygen available and the thickness of the protective char layer that will have been generated. The thicker the layer the greater the reduction in the amount of heat generated by the fire.

Understandably the fire service and insurance companies have concerns around engineered timber. The combined use of engineered timber panels for walls, floors and ceilings delivers a significant surface area of combustible material, which needs to be made fire safe. Guidance is available to provide options to match the design solution of the engineered-timber frame to the likely consequence of fire spread in the building and the timber industry now has the tools to consider how engineered timber will behave in the event of a fire.

Tall engineered-timber buildings, like all tall buildings, present specific challenges to fire fighters if a fire does take hold. One such challenge is that beyond the height of the tallest available ladder the fire can only be fought from inside the building. As a consequence engineered-timber buildings need to deliver structural safety over a longer period and higher fire resistance than timber-framed buildings.

Architects were amongst the first to grasp that timber frame and engineered timber are significantly different products in the context of fire safety, in part because engineered timber enables construction to greater heights, which generates more demanding safety issues and which need different solutions. Hence the timber industry, including the manufacturers of engineered-timber products and their trade representatives, have been investing significantly in research into understanding how engineered-timber buildings will perform in the event of a fire, including safe levels of structural strength in a fire, with a view to providing solutions to protect the timber from open fire spread. Organizations like the UK's Structural Timber Association[8] and Timber Development UK[9] are leading the way in guidance and research on fire safe solutions for engineered-timber builds.

Currently two main fire protection solutions exist. The first option is to reduce the risk of fire spread from exposed wood surfaces by coating the engineered timber with a flame-retardant treatment (FTA) as part of its manufacturing process.[10] As Timber Development UK point out "flame retardant chemical treatments for timber work by reducing its ignitability, slowing down the surface spread of flame and, as a result, reducing the heat released. This allows more time for escape in a fire situation. These effects can improve the material's 'reaction to fire' classification significantly".[11] The second option is to cover the engineered-timber surfaces with a protective layer of non-combustible material such as dry lining to stop fire reaching the timber. This was the approach used in the construction of Tallwood House in Vancouver (as noted in Chapter 4).

An additional factor to consider is how engineered timber responds if it does come into contact with fire. If a fire were to break through the protective layers introduced by either a fire-retardant treatment or by a protective non-combustible material then the consequences might be reduced by the intrinsic fire-resistant properties of large solid wood. When timber begins to burn the surface begins to char. The charred surface is the result of the chemical change in the wood that the fire has caused. The char creates a thermal protection surface over the unburnt timber below the char depth. This protection *delays* the onset of the heating of the core of the wood within the timber section at a depth below the char layer. Consequently, charring is often viewed as providing timber with an inbuilt natural defence against fire. This is what happened when the Cooperage caught fire as outlined earlier in the chapter. The fire resistance of timber, or the time until the fire has penetrated the material, is heavily dependent on the char layer formed. If the fire is allowed to develop – to above 100°C – there is a heat zone in the timber, which can be 10–20mm below the charred surface, where the wood has no strength until the temperature drops to below 100°C. After a fire event the heat that is now trapped behind the char layer can migrate through the timber reducing the strength until that heat is reduced. Char rates are relatively constant for a given depth of material but increased temperatures in a fire can cause increased charring. Engineered-timber structures that have exposed wall surfaces – the panels forming the whole wall not just the frame – will have more fire fuel and that in turn is likely to cause more burning of the engineered timber. However as the timber burns it will char as outlined above creating a thermal protection surface over the unburnt timber below the char depth: "To achieve the designed fire resistance period there must be sufficient virgin solid timber remaining behind the char layer to sustain the loads applied. Therefore each CLT panel within the building must be designed for the fire resistance period and the specific loadings applied to that panel".[12] When the fire service arrive and their actions begin to reduce the heat of the fire the protection provided by the char increases and the threat to the structure starts to decline.

To summarize, the three different types of timber builds discussed here – the different typologies[13] – all have different fire safety considerations. In the view of the Structural Timber Association's technical adviser Martin Milner, when it comes to fire safety, it is not about the material. It is about considering the design and matching safety considerations to specific needs. His message is not that engineered timber is more dangerous in a fire. The important thing, he argues, is for the design team to resolve the new issues which timber presents. If this is done, the result will be a fire-safe, sustainable and likely energy-efficient building. All structural mass-timber buildings should be assessed on a case-by-case basis to ensure that the individual rooms and the building, as a whole, will be safe for occupants. With non-combustible materials, the review assessment is less onerous, but the end result of a properly designed and assessed timber-based building will be the same. In the final analysis, Milner argues, structural mass-timber buildings can be designed and built that deliver fire safety for their users. The benefits of a natural building product can be completely realized with the right fire safety strategy. This is not about one-size-fits-all; it is about engineered solutions using known, modern and proven science applied to sustainable buildings. There are different solutions for the different requirements relating to the shape, size and use of the buildings concerned.[14]

Similarly, Birgit Östman, a senior fire researcher at Linnaeus University, Sweden, who has worked on construction and fire for the past 30 years, is convinced timber used to form the structure of a building is no more dangerous than concrete or steel: "The fact is that wood has many fire safety benefits compared with other materials. Take steel, for example, which certainly doesn't burn, but it does soften and lose its structural integrity. Although wood is flammable, it burns in a controlled and predictable way. Wood also retains its load-bearing capacity for a long time, even during a fire".[15] This predictability, an attribute often associated with wood in relation to fire, is however predicated on knowing the temperature of the fire.

While Finland and Sweden are the two most afforested countries in Europe you would until recently have struggled to find a major wooden building in their respective capitals, Helsinki and Stockholm. This was due to nineteenth-century fire regulations that prohibited building tall in timber. However, since both joining the EU in 1995 these bans have been lifted and they adhere to the EU's Construction Products Regulation, which contains seven core requirements of which fire safety is one. In the event of fire, the key principles are: residents must be able to leave the building safely; structural elements of the building must maintain their function for a set time; the development and spread of fire and smoke must be restricted; the spread of fire to adjacent buildings must be restricted;

and the safety of emergency service personnel must be taken into account. Engineered timber buildings can meet all these requirements.

Of all the European capitals it is London that has the largest number of big new engineered timber buildings – surprising given how little forest cover the UK has. These timber buildings are concentrated in the Borough of Hackney as a result of the Council's planning policy which seeks to "incorporate sustainable design and construction measures and materials, demonstrating how sustainable development principles and resilient measures to climate change have been incorporated". This has been incorrectly shorthanded into a "wood first policy". It is not. The policy does not preclude other construction materials. However, by stressing "sustainability" and "climate change" it has consequently encouraged the use of wood as a primary construction material. Hackney now has 24 engineered-timber buildings comprising seven schools, six mixed-use including a cinema, ten residential buildings and one church. This is the highest concentration of new large engineered-timber buildings in the world.

However, the tragic Grenfell Tower fire in London, which claimed 72 lives in 2017, has – quite rightly – drawn attention to the dangers of fire in tall buildings. Controversially it has proved to be a major setback for building tall and big residential buildings with wood in the UK. Despite Grenfell not being a timber building (structurally it was made from reinforced concrete clad in polyethylene, a form of plastic sandwiched between two very thin sheets of aluminium) the capital's politicians, along with central government, have introduced a ban on any combustible *structural* material being used in new buildings over 18 metres tall – or to be precise 18 metres from the ground to the floor of the top storey. In addition, for timber to be used over 11 metres and up to 18 metres a large-scale fire test has to be passed, which acts as a major disincentive to go above 11 metres. The 18 metres is dictated by the length of the longest ladders held by the London Fire Brigade – basic and brutal. Perhaps consideration should be given to fire and rescue services in all major cities, irrespective of whether they have any tall timber buildings, having at their disposal the longest ladders and the tallest aerial platforms. Grenfell Tower in London is 67 metres tall and there are commercially available skylifts that can reach beyond 70 metres.[16]

In the wake of Grenfell, some may think the 18-metre height ban in the UK understandable but, as previously noted, it is the *contents* of buildings where most fires start and develop. The logic really should be to ban combustible contents but this is a non-starter as a quick look round any home will tell you. The ban on using timber as a structural material in buildings above 18 metres *because it is combustible* is simply wrongheaded. Current research has shown that you can build above 18 metres safely and there is no evidence that engineered timber used to form the structure of a building is more problematic than any other material when fire breaks out. Such a ban can only make sense if it applied to constructing any building over 18 metres from any structural material. Having said that anyone involved in constructing tall buildings irrespective of what structural material they use should read Peter Apps's harrowing account of the Grenfell fire so that they are aware of the potential dangers inherent in *all* tall buildings anywhere in the world.[17]

One change in the building regulations in the UK that has been widely accepted post-Grenfell, including by the wood industry, is a ban on virtually all combustible *cladding* materials. As a consequence, materials such as plastic, timber and high-pressure laminates are no longer allowed on buildings over 18 metres as of 2022. Cladding is the material used to cover the exterior of a building, not to provide its structural strength. Initially in 2018 the ban was proposed at 11 metres, but this was then changed to 18 metres with the rider that the applicant had to demonstrate that this was workable. This has proved to be a hurdle too high for many resulting in 11 metres becoming the de facto limit, as it is for engineered timber builds.

While London has reduced the maximum height of residential timber buildings can be built to, other countries are going in the opposite direction. Denmark has lifted the limit to 45 metres and Canada to 42 metres. Paris is now pulling ahead of London when it comes to building in wood.

With half an eye on good PR for the 2024 Olympics France has pledged to build more with wood. Indeed French government policy dictates that new public buildings must be made from at least 50 per cent timber or other natural materials from 2022. The eagle-eyed reader will have spotted that, in Chapter 4, Tallwood House in Vancouver was listed as being 54 metres tall, which is 12 metres higher than permitted by Canada's National Building Code. How was this possible? Tallwood House was deliberately built on native American land where federal government regulations do not apply!

Globally the wood industry continues to invest in research into fire and wood and, as each year goes by, a growing body of scientific evidence supports the notion that, *designed and built correctly*, large wooden buildings are just as safe as those made from any other material.[18] If we focus in on Mjøstårnet in Norway, currently the world's tallest timber building, we see that the following fire safety measures have been implemented, as described by Green Building:

1. The glulam structures: The glulam structures have such huge dimensions that they retain their load-bearing ability in the event of a burnout fire. The glulam structures are positioned such that they won't affect each other in the event of a fire.
2. Sprinkler system: A state-of-the-art sprinkler system covers the building inside and out. It differs from a traditional sprinkler system in that it is scaled to provide a greater water volume, it has an independent extra water supply, as well as a greater degree of monitoring.
3. Facilitation for the fire brigade: Dedicated control room with graphical overview that enables fast and efficient extinguishing. The fire alarm is linked directly to the fire brigade, and the fire brigade's water supply is doubled.
4. Fire strips: The weakest point in a timber structure in the event of fire is the steel used for the actual connections in the structure. The glulam structure therefore has fire strips to protect the steel

sheets and dowels in the junctions and joints. The fire strips consist of a material that expands 20 times at 150 degrees Celsius. The strips protect the steel from the increase in temperature, close openings and prevent fire from spreading.

5. Fire cells: Each individual floor, apartment, and each hotel room are designed as separate fire cells that will restrict fire from spreading [Some concrete has been used to create these fire walls].
6. Protection against fire spreading in the façade [cladding]: The outer wall elements have been treated with a fire-retardant material, and cavities in the façade are broken on each floor.[19]

Observing the wood and fire risk more generally, it is the case that fires have broken out on building sites while timber buildings are under construction, but this is more often down to poor site discipline linked to workers having little or no experience of building with wood. Once the workforce become familiar with working with timber and the necessary guidelines are followed such incidents become much less likely. When construction is completed and the building is occupied the risk of fire drops appreciably.

It should also be noted that experience from both Switzerland and Sweden shows that quality control can be made much more efficient when building in wood since wooden buildings are industrially produced (i.e. in a factory setting) to a greater extent than concrete or steel frame buildings, hence quality control procedures can be implemented inside a controlled production process (see Chapter 11). This greatly enhances the possibility of ensuring that fire safety measures are undertaken properly.

To date engineered timber construction and insurance have had a tricky relationship, but there is growing support and appetite within the insurance industry to insure more engineered timber projects. Insurers have unhelpfully lumped together all timber in the same log basket. They have failed to understand that there are different construction technologies being deployed here and consequently it is wrong to compare a tall, big, solid engineered-timber build with a three-storey timber-framed house.[20] Such comparisons, equating engineered timber with the more traditional timber-frame build, are leading insurers to make false assumptions about timber construction. The first of these is that a fire in an engineered timber building automatically leads to the total loss of value from that asset since the whole structure inevitably burns down to nothing but a pile of ash. This is simply not the case. Engineered timber buildings can be designed to accommodate hazards without leading to extensive loss. If there has been a fire or flood timber is pliable so parts of the building can relatively easily be replaced which is not always the case with steel and concrete buildings.

One of the most common problems associated with insuring an engineered-timber building is failing to put in place risk mitigation measures early on when designing and constructing an engineered timber building. This limits the time an underwriter has to perform the necessary due diligence. A positive way to alleviate the concerns of insurers is to invite them on a timber discovery journey at the earliest possible stage. Kirsten Haggart, Senior Associate at Waugh Thistleton Architects, believes: "We need to understand the concerns from an insurance perspective. As architects working with timber, we are embedded in the process, we know what the risks are and how to mitigate them in our design. We forget that others aren't in that world yet and it's important we share our concerns and our knowledge so that risks are seen realistically".[21]

Addressing the concerns of insurers with regards timber buildings as early as possible is also the advice of Philip Callow, the founder of Mass Timber Risk Consulting. Callow's view is that this places a bigger burden of responsibility on the design team hence he suggests creating a report for underwriters that: ". . . articulates the challenges, articulates the fact that you recognize them as hazards and that you have mitigation plans for them, including water management, buildability, what happens during construction, as well as replacement and refurbishment. And this must be done in an easily digestible way to ensure it is simple for the insurers to understand".[22] The risk matrices used by insurers are based on historical evidence and they often lack data about engineered timber buildings. This places the onus on those with knowledge to help fill these gaps. According to Callow: "There are unknowns in an area where we don't have enough data. If architects and developers can answer these questions and demonstrate through design and risk management why their building is better than a normal building then they should get an equitable response".[23]

A further option is to develop a databank of examples that can be shared. The organization Built by Nature, working alongside the University of London, is taking a major step in providing this evidence.[24] They are devising a set of pre-warrantied models for residential buildings that have been approved by inspectors and insurers. These will allow any architect or contractor to build from these designs with the knowledge that these buildings can be warrantied, will be approved by the fire service and, most importantly, can be insured.

As well as sharing best practice, much can also be learned by passing on information about failures. While there are inevitably sensitivities about sharing such data the risk of perpetuating previous mistakes could also create a long-term setback for timber buildings and cast an indelible stain on this much-needed evolution in building practices. Collective knowledge can play a big part in breaking down these barriers as well as early engagement with all stakeholders. At the same time associated industries – such as insurance and investment – need to be open-minded and willing to embrace this necessary change particularly as this timber transition is not just about reducing emissions and storing carbon – crucial though they are – but also about building more cost-effective, high-quality buildings with better working and living conditions for people.

This chapter has focused on fire safety in relation to the wood used to form the structure of a building. Increasingly, as we discussed in Chapter 3, wood will be used to manufacture insulation. Some types of insulation will burn in a fire but when it comes to wood fibre there is the following reassurance from Back to Earth, a supplier of sustainable building materials:

> Natural fibre insulation materials tend to burn slowly and give off fairly benign levels of smoke. Whilst they will burn in a fire they do not contribute to the fire as they tend not to support the flame. In the case of wood fibre insulations, the rigid boards are dense, contain little oxygen and smoulder slowly, taking many hours to burn through. The best analogy would be during a book. Putting it page by page on the fire it will burn quickly, however, a whole book in a fire takes many hours to burn through.[25]

In summary, the timber industry recognizes that to increase the use of sustainable wood in construction requires answers to fire safety questions. The timber industry takes fire seriously and has put significant investment into establishing how the use of timber can be incorporated to create a fire-safe building and is leading the way on fire safe building technology. Fire safety in historic post-and-beam and modern timber-frame buildings is about recognizing that the fire fuel is low and that the wood is protected from room fires by dry lining. In the event of a fire the fire brigade will have time to intervene. Engineered-timber fire safety is about recognizing that the fire fuel is high. There are then three conditions to consider.

1. If the product is exposed for aesthetic effects, it will require an engineered review to show how it would behave in a fire and be designed to burn itself out by the formation of charred surfaces and reduced fire fuel surfaces.
2. The fire from a room's contents must be stopped from reaching the engineered timber, which can be achieved by using multiple layers of dry lining (as per timber frame).
3. If no fire service intervention occurs, a fire safety strategy would be agreed that the building would burn until it has exhausted the available wood as fuel.

While accepting that fire is a formidable force and can exploit any weakness, if a fire does take hold in an engineered-timber building then successfully fighting it will in large measure depend on those fighting the fire understanding how engineered timber performs in a fire and having a thought through and prepared fire-fighting strategy to match this understanding.

6

Forests: an important asset in tackling climate breakdown

Fact: *The European Union's forest area has increased by 10 per cent over the last 30 years.*[1]

As George Orwell noted, "To see what is in front of one's nose needs a constant struggle", hence it is not surprising that although timber offers a significant solution to climate change very few people realize this. As I have argued, timber is the material we must use to replace our main carbon-intensive construction materials, especially concrete and steel. It can both substitute for these materials and safely store carbon. To help us understand timber, and why it has such potential to help us, we should visit the forest. This book is pro-timber, but also very much pro-forest.

Almost a third of the world's land surface is covered by forests.[2] Europe's forest cover is 35 per cent – over a third of the continent.[3] UK forest cover stands at 13 per cent.[4] Within the EU Finland (74 per cent of total land area) and Sweden (69 per cent) are the most heavily forested countries, followed by Slovenia (61 per cent), Estonia (57 per cent) and Latvia (55 per cent). The Netherlands (11 per cent), Ireland (11 per cent) and Malta (1 per cent) are the least forested countries.[5] If we consider things at a European level – which brings in the Russian Federation, Norway, Switzerland, Ukraine, Belarus, etc. – then forest cover is around 1,000 million hectares, which is about 30 times the size of Poland (see Figure 6.1). The largest forest area in this wider geography is located in the Russian Federation.

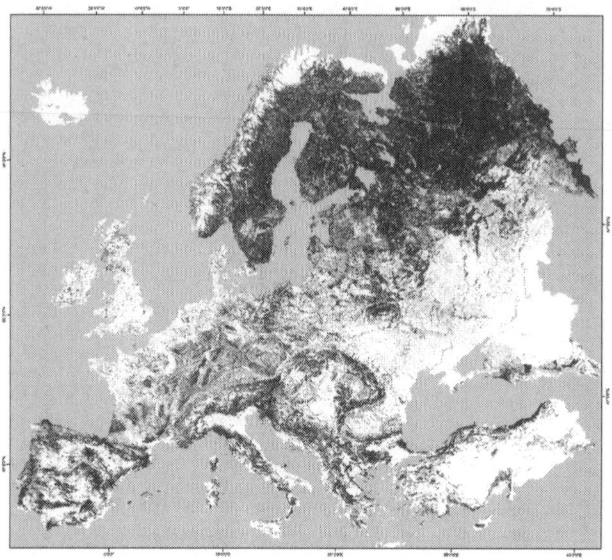

Figure 6.1 Forest Map of Europe
Source: European Forest Institute/EC Joint Research Centre.

All trees – big or small, deciduous[6] or coniferous[7], on their own or with other trees in a forest – have the ability to absorb carbon dioxide (CO_2) from the atmosphere and exhale oxygen in return, which is why forests are often rightly referred to as "the lungs of the earth". However, as Harald Mauser of the European Forest Institute points out, "This is reversed when trees are dead and decay. This releases almost all the CO_2 they stored except a very small part that was transferred to the soil via litter and roots".[8] The EU's forests absorb the equivalent of nearly a tenth of the EU's total greenhouse gas emissions, including CO_2, each year.[9] This is impressive in itself, but also hints at the huge potential offered by forests and the timber they produce.

While expanding forest cover in Finland and Sweden may not be an option, as they are already heavily afforested, there are plenty of other European countries which could increase their cover. Romania and Bulgaria stand out, but France and even the UK could notably increase their forest cover if the political will was there. To state the obvious, more tree cover (more planting) would mean more carbon absorbed (sequestered) out of the atmosphere and stored safely in the wood. But it is not quite that simple. If we plant a forest and then simply leave it unmanaged it will not absorb as much CO_2 as a managed forest. Managing forests includes such practices as cutting back undergrowth to allow trees to grow more easily and thinning out the number of trees, which enables those remaining to grow bigger and faster *and overall absorb more* CO_2. As Giffen *et al.* note: "Improved forest management can lead to substantially increased carbon storage simultaneous with increased timber harvests, which allow for additional carbon storage in harvested wood products and reduced GHG (Green House Gas) emissions from substituting wood for more CO_2-emission-intensive materials".[10]

We need to manage more forests in Europe to maximize their carbon storage as well as other services we request from them, such as timber production. In the UK 44 per cent of our forest cover is actively managed but with some big national disparities. In Scotland it is 60 per cent but in England a mere 24 per cent.[11] Not surprisingly commercial conifer forest is managed whereas non-commercial mixed or deciduous woodland is often left to its own devices. The Royal Forestry Society (RFS) believes that there should be a target for managed forestry in England and Wales of 75 per cent: "If 75% of woodland was in active management, the RFS conservatively estimates the additional timber produced would boost the economy by £20m and create up to 240 rural jobs across England and Wales. Contractors and processors would also benefit significantly from increased timber supplies".[12] While the RFS focus here is on increasing timber production such a target would also increase the amount of stored carbon.

The way forward is to both increase forest cover in rural areas (where most forests are situated) and increase the amount of forest that is actively managed. There is also plenty of scope for smaller-scale planting, especially in and around our towns and cities. This would help to generate many benefits in addition to storing more carbon. The shade created by trees would cool our urban environments in the summer. Our cities would be more aesthetically pleasing. And we would have more biodiversity, reduced run-off into our drains (tree roots guide water underground and then "drink" it up) and better air quality thanks to the trees sequestering carbon dioxide, capturing pollutants and producing oxygen. Next time you move around an urban area start looking to see where you could plant trees: four on that piece of grass, a line down the side of that road, a small clump at the top corner of the school field. It all adds up. What we should avoid at all costs is deforestation. Thankfully this is not a widespread European practice except when new roads or railway lines are built or cities and towns expanded.

Although Europe's forest cover has expanded in recent years, deforestation[13] in the Global South, for instance in Papua New Guinea and Brazil, is on an immense scale.[14] Swathes of forest are being felled with the trees being removed and *not replaced*. A note of caution here. While Europe's forest cover may be expanding it is our apparently insatiable demand for the likes of beef, soya and palm oil that is helping drive deforestation globally.

In Europe to cover the cost of land ownership, forest management and adapting forests to climate change, a section of the forest will be felled for commercial use but it will almost certainly be replanted within 1–5 years,[15] with two or three saplings planted for every tree felled. There are occasions when the forest is not replanted but often for good reason – for example if the trees were to be planted on deep peat,[16] which today we would avoid as such planting releases more carbon from the peat than the tree will absorb. In my nearest large forest, Kielder in Northumberland, several clear-felled sites in recent years have not been replanted, allowing the deep peat to rejuvenate and re-establish and do what it does so well, store carbon.[17]

Harvesting trees (chopping them down) is regulated by legislation in all European countries. Harvesting can be done by felling single trees, groups of trees or typically by clear-cutting a complete forest stand. While in the past in some regions clear-cuts may have been upwards of 100 hectares, today in most European countries their area is limited by law, for example, five hectares in Sweden, two hectares in Austria. A larger clear-cut may be carried out as a result of storm or pest damage.

When large clear-cuts were introduced in Sweden in the 1950s the spirit of the time was efficiency and mechanization and the aim was to restore the Swedish forest for wood production after its condition had become degraded at the beginning of the century. But the clear-cuts were very large, lacked environmental consideration and often included the widespread use of pesticides, leading, in the 1970s and 1980s, to protests against large clear-cuts. Sweden introduced new forest legislation in the 1990s which gave equal weight to timber production and the environment. Now legislation and certification require environmental considerations to be taken into account in all forestry operations, which has meant avoiding clear-cut areas that are too large, allowing "nature-value" trees (often old coniferous trees or broad-leaved trees with particular bark characteristics) to be left in place, along with trees located near watersheds and other sensitive biotopes (habitats associated with a particular ecological communities). Regulation also requires that deadwood be preserved to promote wood-living insects.

Wondering if there was an optimum area size for a clear-cut, I raised the question with Magnus Berg of the Swedish Forest Industries Federation:

> I would say that there is no one optimum size for a clear-cut that is true for a whole country as the conditions vary. The size will depend on a number of factors, such as site ecology, forest condition, forest history, forest damage, economic factors, etc. But it is true that with a smaller clear-cut, it will be more expensive to do the harvesting operations and you will also need to build more forest roads and have increased transport.[18]

Environmental NGOs can get rather exercised by clear-cuts and have a tendency – in the view of the forest industry – to overstate their environmental impact on biodiversity including wildlife. In the UK a mosaic of felled areas provides valuable habitat for nightjar and woodlark. By way of comparison, it should be noted that farming entails *annual* clear-cuts of cereals or vegetables, as opposed to every 30–100 years in the forest (depending on tree species and growing conditions), so it could be argued their cumulative environmental impact is greater.

One hundred years ago UK forest cover hit a low point of 5 per cent of its landmass. Since then it has increased but overall UK forest cover remains low by European standards at only 13 per cent compared to around 40 per cent for EU countries, 35 per cent for Europe and 30 per cent for the landmass globally. Although within the UK only 2.5 per cent of its landmass is ancient woodland/old growth (i.e part of the 13 per cent), which is a precious pool of biodiversity and an important genetic databank that we would be fools not to protect.[19]

The UK's commercial conifer tree forests are often criticized for lacking biodiversity. A hectare of *commercial* conifer forest, even if it was planted 100 years ago, will likely have lower levels of biodiversity than a hectare of native forest of the same age. However, that does not mean there is no biodiversity. In 2020 Confor, the UK Confederation of Forest Industries, based in Edinburgh, produced the report, "Biodiversity, Forestry and Wood".[20] In the report, author Eleanor Harris points out that mature forestry plantations can provide outstanding biodiversity habitat. She argues that the structure of the woodland is more important than the species of the tree, and well-managed, commercial conifer forests can provide the light levels, scale, structure, dynamic change, and other conditions required by forest wildlife.

Confor would be the first to acknowledge that the monocultural planting of conifers in the 1960s and 1970s did not support much diversity and, more importantly, damaged some existing habitats which were important in their own right. However, the challenge of producing timber while at the same time boosting biodiversity is one commercial forestry can meet. Every parcel of forest, known as a stand, felled presents us with a new set of possibilities: we might simply replant like for like, or maybe the felled trees were planted on peat and now is the opportunity to restore the peat, maybe the felled trees were all of the same species and now is the opportunity to plant more of a mix, maybe the felled trees ran right up to the edge of a stream and now is the opportunity to redraw the tree line. Such interventions can help increase biodiversity with little or no impact on levels of timber production. Mixed planting also reduces the risk of disease ripping through a whole forest – a good tree species to plant today may prove to be not such a good choice in 40 years' time.

In the UK there is widespread, cross-party political support for significantly increasing UK forest cover. Scotland currently has 19 per cent cover compared to Wales with 15 per cent, England 10 per cent and Northern Ireland 9 per cent.[21] These figures are too low in the opinion of the independent Climate Change Committee who have recommended at least 17 per cent for the UK as a whole by 2050.[22] The most recent figures are not reassuring. The UK planted a total of 12,960 hectares of new woodland in the year to 31 March 2023 – less than half the Conservative government's manifesto target of 30,000 hectares of new woodland creation annually by 2025 and down from 13,860 hectares in 2021/22. Behind the UK headline figure in the period 2021/22–2022/23, England saw an increase from 2,260 to 3,130 hectares, Wales doubled from 580 hectares to 1,190 hectares, Northern Ireland slipped back from 540 hectares to 450 and in Scotland new planting dipped from 10,480 hectares to 8,190 hectares in the same period.[23] If the government were to deliver on its annual target then this would just about ensure the 2050 target of 17 per cent tree cover was reached.

However, it is surprisingly difficult to plant a new forest in England, *especially* a commercial timber forest (and it's not exactly easy in Scotland or Wales either). From 2008 to 2018 no new substantial commercial forest was planted in England. In 2018 a new commercial forest began to be planted at Doddington, in north Northumberland, close to the border with Scotland, but nothing of any size since. Native forest planting – not for timber production – has proved less difficult, for example the Woodland Trust's Heartwood Forest in Hertfordshire and the slowly advancing Northern Forest, a mosaic of woods straddling across northern England from Liverpool to Hull.[24]

Doddington is 354 hectares[25] and the land has been allocated as follows: Sitka spruce 41 per cent, native broadleaves 20 per cent, managed priority habitat 15 per cent (leaving land as it is and not planting on it because it provides a type of wildlife habitat that is under threat), Scots pine/native broadleaves 13 per cent, open ground 10 per cent and mixed conifer 1 per cent. Please note that Doddington has been planted as a commercial forest, meaning that its primary purpose is to produce timber, despite containing a third tree species native to the UK which are being grown first and foremost for their biodiversity, not their timber.

Planting Doddington, however, was not an easy task. It needed the support of both the local MP and MEP to push through approval of the project in the face of resistance from the statutory stakeholder community. The early indications from Doddington are good. Six years since the first planting, the mixed tree landscape is establishing well and quickly. Notable successes include the oaks which are several feet tall already, as a result of being encouraged to compete with birch and alder through planting density and species combinations. Moreover, the mixed woodland has seen biodiversity return, including numerous raptors, invertebrates and soaring numbers of passerine birds.[26]

Elsewhere in the UK new native forests are being planted with a preponderance of deciduous trees, often referred to as broadleaf. One of the largest planting schemes in England is the Heart of England Forest, known to the locals as "The Forest of Dennis" after its benefactor, the mercurial publisher and poet Felix Dennis.[27] Having become a multi-millionaire Dennis decided to "give something back" and began buying up large tracts of prime farmland around his Dorsington residence in Warwickshire – much to the irritation of the National Farmers' Union. Dennis has said his work was necessary because the UK was so denuded of native trees: "France, Germany and Italy have approximately 25% of native high-tree cover. In parts of England, we're lucky if we've got 1.5%. Britain has the lowest percentage of native adult trees of any country in Europe except for Ireland and Iceland. We're so far down the league table you can't even see us".[28] The Forest now covers a mosaic of habitats across 2,800 hectares of land, including 1,800 hectares of new woodland and 240 hectares of beautiful, mature and ancient woodland. In 2022 the 2 millionth tree was planted.

The focus on "native" species does raise some questions. According to the Royal Horticultural Society, "It is widely accepted that 'native' trees and shrubs are those species that have occurred naturally in Britain since the last Ice Age". This definition rules out a surprisingly long list of widely perceived and much-loved British trees. For instance, the sycamore tree including the one that was the star of Sycamore Gap on Hadrian's Wall before being inexplicably felled in the dead of night in 2023 – is technically an interloper that Dennis's strict planting regime prohibits.[29] The sycamore may have arrived with the Romans or in the 1500s: either way it is not strictly native, as too are the apple, copper beech, cherry plum, sour cherry, field elm, horse chestnut, sweet chestnut, holm oak, pear, London plane, plum, walnut and Irish yew.[30] In my experience "native trees" is no more than shorthand for "broadleaf trees I like" and hence "non-native" means "trees I don't like", especially commercial conifers!

That said if we can double forest cover in the UK while having a roughly equal split between commercial trees for timber and broadleaf trees for biodiversity, most parties will be satisfied. Even non-native, non-commercial timber trees should do just fine, partly because we will continue to plant them in our gardens.

That the UK needs more commercial forestry is reflected in the figures for timber imports that show the UK to be the third biggest net importer of timber after China and the USA.[31] As it stands the UK imports 80 per cent of its wood,[32] a much bigger figure than our imported food at 46 per cent.[33] Such offshoring of both wood and food production means that large tracts of land elsewhere in the world are being used to feed the UK's consumption (see Figure 6.2).

The problem at the moment is that we are not planting anywhere near enough trees in the UK to meet the target of doubling our forest cover. The less we plant the greater the arguments will be about what we plant and for what purpose. Instead, we need to grow the supply so that everyone can access enough for their needs. But how fast is it growing?

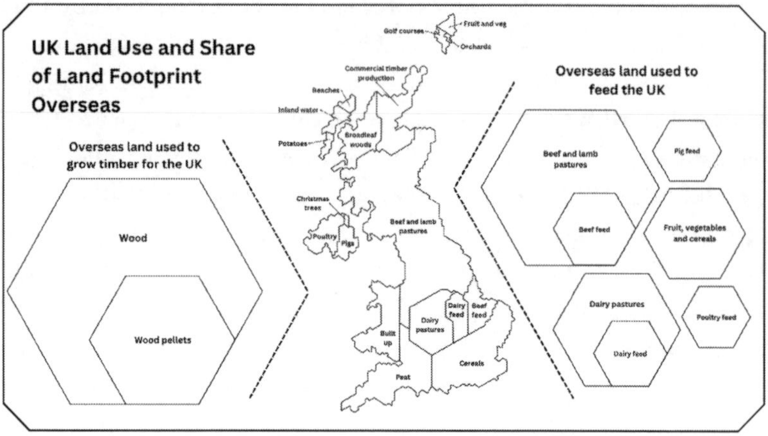

Figure 6.2 UK land use and share of land footprint overseas to meet the UK's timber and food needs

Source: Confor.

Of the 13,860 hectares of new woodland of all types planted in the UK in 2021–22 76 per cent was planted in Scotland, 16 per cent in England and 4 per cent in both Wales and Northern Ireland.[34] More broadleaf trees were planted than conifers across the UK with only Scotland putting more of the latter in the ground. The Sunak Conservative government pledged to plant 30,000 hectares of new woodland every year from 2025[35] meaning that the 2021–22 planting figure is less than half of what will be needed. Needless to say, the forest community is not happy. "This is a total policy failure in both economic and environmental terms", stated Stuart Goodall, Chief Executive of Confor. Goodall continued:

> Report after report has shown that increased tree planting and wood use is vital to meeting the UK's net zero targets – yet this is not being translated into trees in the ground. This is especially true of productive forests, which produce wood alongside benefits for carbon, nature and the economy. The Government knows more wood is needed to build low-carbon homes and it knows the UK

> faces a shortfall in future domestic wood supply, yet it is seeking to keep a tight lid on the planting of productive forests. Environmental groups know we need to grow more wood in the UK rather than increasing the pressure on fragile global forests, but the Government – against all the evidence – is not taking decisive action.[36]

There is a well-established myth that timber grown in the UK is not of a high enough quality to be used as sawnwood in most construction applications. That this perception exists is understandable – as I shall explain – but that does not make it true. Timber from Sweden, the UK's leading sawnwood supplier, and from elsewhere in Europe, is exposed to different growing conditions, including longer rotation lengths,[37] to conifer forests in the UK. Consequently that leads to most of their sawnwood meeting the higher grade C24, as opposed to UK spruce, which has shorter rotation lengths and mainly grades out at C16, although it does produce some C24 (the C standing for conifer).[38] Currently only 27 per cent of UK sawnwood is used in construction because imported sawnwood, led by Sweden, is seen as being of a higher quality – and it is.[39] The bulk of UK timber is used for fencing (39 per cent) and packaging and pallets (25 per cent).[40] In reality, testing has shown that UK C16 timber is *good enough* to undertake 90 per cent of construction applications and it is used by some mainstream timber frame companies, however, structural engineers who advise architects and others who commission construction projects often persist in specifying C24. This risk aversion has resulted in "over engineering", as with steel beams (see Chapter 1) and given C16 sells at a lower price than C24 this means the construction cost is higher than necessary with UK timber unfairly discriminated against. This situation needs to change.

At this point, let us introduce a new contributor to the argument, the Sitka spruce (see Figure 6.3). The Sitka spruce is an alien to the UK having arrived in 1831. Its ancestors originated from the west coast of America. Following the Second World War the UK planted millions of Sitka spruce,

Figure 6.3 The much-maligned Sitka spruce
Source: Shutterstock.

often in great swathes of monoculture – something that has been rightly criticized and which is no longer done. But although much-maligned, given the climate and latitude of the UK, no tree can grow faster, and no tree can absorb more carbon dioxide while at the same time producing good-quality commercial timber. Today approximately 50 per cent of the UK's commercial planting comprises Sitka spruce.

Recent government research in the UK by Wilkinson, Xenakis and Morison compared the carbon balances of two forest stands: an upland Sitka spruce plantation in Harwood Forest in the northeast of England and a lowland deciduous oak plantation in Alice Holt Forest in southeast England. Their research found that:

> Overall total carbon stocks, including tree stems, roots, litter and soils, were higher at the spruce forest compared with the oak forest and soils at the spruce forest contained a larger proportion of the total carbon (62%) compared with the oak site (49%). The evergreen spruce forest remained photosynthetically active all year round, including over the winter period when the deciduous oak forest was leafless. The average overall long-term net ecosystem carbon dioxide uptake was 53% higher at the spruce forest compared with the oak forest.[41]

Given that we are knee deep in a climate crisis there is a great deal to be said for planting a tree species that can sequester and store a lot of carbon – and fast. A Sitka planted today will be reaching maturity and be ready for felling in around 30–35 years' time, by which point we will either have succeeded in avoiding climate breakdown or not. If we have stabilized the climate by 2050 then we can think again about our replanting options. Having planted Sitka now does not mean we have to plant Sitka again in 2060. Rather we can revise our planting ratios which may involve increasing the native mix to a level greater than we have done, say at Doddington, or the evidence may point towards planting a lot more of something that was not even in the Doddington mix – palm trees perhaps, if we have wrecked the climate. Commercial tree planting is not irreversible. After felling we can decide to replant differently. However, let us recognize that in the UK the Sitka is the premier tree in our battle against climate change. Let us stop maligning it and instead let us continue planting a significant amount of it. In reality it is a climate champion.

The biodiversity which Sitka supports may come as a surprise to the many who have bought into the "native deciduous good, foreign conifer bad" argument. Kielder Forest, the UK's largest planted forest, mainly in Northumberland, is 75 per cent Sitka and is one of the last remaining strongholds of the indigenous red squirrel in England – home for over 50 per cent of the population and maybe as high as 75 per cent. The reason for their success is that the Sitka spruce is not suitable for the red squirrel's competitor – the invasive *grey* squirrel.[42] It is not only the red squirrels who like Sitka. The dense foliage provides cover from the wind and rain for larger mammals while birds of prey and smaller birds, such as crossbill, tree creeper, coal tit and siskin, may use Sitka spruce for nesting. If the lynx were ever to be reintroduced our conifer forests with their high percentage of Sitka would be the ideal territory.

I hold a life membership of the Woodland Trust, the UK's largest woodland conservation charity. I think they do great work and, if we were to double UK forest cover, I would be keen to see them playing a key role in managing our native woodland. However, I do get a little narked that the importance of quickly capturing carbon dioxide out of the atmosphere is not part of their message. The oak saplings that they freely distribute will over time become fantastic for biodiversity but they grow very slowly. By the time they reach their maximum carbon sequestering potential they will be at least 100 years old, at which point we may well have lost the climate change battle. My challenge to the Woodland Trust would be to encourage people to plant not one but two trees in their garden. Yes, please plant a native deciduous tree – lots of benefits especially for biodiversity – but please also plant a second quick-growing conifer, like the Sitka spruce, or a juniper, with the primary purpose of absorbing carbon dioxide from the atmosphere – fast.

We need to better understand the benefits of forests and trees, especially their ability to sequester CO_2 and store carbon. Many more people in recent years have sought to understand where their food comes from. We need to embark on a similar journey regarding where our *wood* comes from, along with a better understanding of the climate benefits that wood offers. This would generate greater public support for the forests and trees we already have and, where appropriate, increasing their number significantly. Such an increase would deliver multiple benefits – most notably boosting the supply of timber, our key material for decarbonizing the built environment.

7

Is there enough sustainable wood?

Fact: *The World Bank estimates that global timber demand is set to quadruple by 2050. Demand in the UK is also expected to rise, in part because of the government's commitment to promote timber use in construction as part of the UK's net zero strategy.*[1]

Using more wood in construction and renovation to turn our built environment from a climate problem into a climate solution is predicated on there being enough wood and wood fibre to do this.

First and foremost, this supply of wood has to be sustainable. For our timber supplies to be deemed sustainable the wood must be sourced from a managed forest where natural regeneration of the forest is viable or where new trees are planted as others are felled. Such replanting needs to see between 5,000–10,000 saplings planted per hectare, replacing around 500 felled trees per hectare. The forest ecosystem is required to be regularly monitored to ensure it remains in good health and able to provide for both people (employment, recreation) and nature (biodiversity) now and into the future. A certification scheme accompanies this process of which more than 50 exist worldwide. Of these, the Forest Stewardship Council (FSC) and the Programme for the Endorsement of Forest Certification (PEFC) are the two best known.[2] While both schemes are committed to the same cause – sustainable forestry – the primary difference between the certifications is their origins. Initially the FSC scheme was developed for tropical environments and not suited for forests in Europe and North America. This led to the introduction of PEFC in the late 1990s to facilitate sustainable forest management (SFM) certification in Europe. PEFC now accounts for over 264 million hectares of certified forests, 10 per cent of the world's forest area, and its certification system is recognized in over 30 countries. Meanwhile FSC has certified forests in over 80 countries with 7 per cent of the world's forest area (180 million hectares) carrying the FSC certification. Fifty per cent of people in the UK recognize the FSC logo in relation to buying wood products.

The good news is that Europe's forests have been growing in size in recent decades not reducing as they did during the sixteenth and seventeenth centuries when forests were felled to make way for agriculture and the timber was needed to make charcoal and build ships. Once deep-mined coal became available the felling of forests for charcoal receded and iron and steel replaced timber for shipbuilding. At this point Europe's forest cover began to stabilize and in recent years it has increased (see Chapter 6).

The amount of timber felled and removed from Europe's forests each year is undertaken in the context of what is referred to as the net annual increment (NAI). This is the amount the forest grows each year, that is in term of how much timber is in the forest, not the footprint of the forest. Each year the trees grow that bit bigger – they put on volume and hence weight – and the forests become denser. In 2021 the average volume of wood per hectare was 169 cubic metres, which is 40 more than in 1990.[3] An important reason for this is increased levels of SFM. At the same time the age structure of forests has altered and the share of older forests has increased by more than that of other types of forest everywhere in Europe. As a result, we have more wood in Europe's forests and we can measure this annual "expansion" as the NAI. (If this year there were 300 million metric tonnes of wood in the UK's forests and last year 270 million then the NAI would be 30 million metric tonnes). By taking figures over a rolling period the NAI can be calculated more accurately as there can be blip years caused by ideal growing weather leading to disproportionately good growth or the opposite: poor growing weather, or pests, disease, fires, storms, all of which might reduce growth within a year. As long as the amount of timber harvested in any given year is less than the NAI then the felling is deemed to be sustainable.

Exactly how much timber is being harvested across Europe is not entirely clear as there are variations in the figures produced by different bodies. Eurostat reports: "In total, an estimated 65 per cent[4] of the net annual increment of wood in EU forests was removed by the logging industry in 2020, the net increment being defined as 'the average annual volume growth of wood less the average (recurring) annual mortality'".[5] According to Eurostat the Nordic countries have a lower NAI (Finland around 60 per cent) than many other countries, including Germany and Austria, although Germany's has been high recently due to the need to remove huge swathes of trees that have been infected with the bark beetle.

A different NAI figure is produced by Forest Europe: "In European forests, every year more wood grows than is harvested. About 73% of the net annual increment is utilized by fellings. The wood increment is higher than in earlier periods. Since 1990 it has increased by approximately 25%. Similarly, the volume of timber harvested has increased steadily since 1990. As timber stocks grow despite the higher volumes harvested, European forests provide a sustainable supply of wood as a renewable resource".[6]

Behind both headline figures – 65 per cent and 73 per cent – major country variations can be found. The Eurostat country figures, with the UK added, are set out in Table 7.1. In 2020 the Czechs, Germans, Dutch and Estonians harvested above 80 per cent of their NAI due to the need for extensive felling to eradicate bark beetle. However, with the exception of the Czech Republic in that particular year, this still means the forest "mass" is growing year on year even though a relatively large amount of wood is being removed. Other countries – such as Italy, Romania, Spain and Bulgaria – make a much lower use of their NAI. This can be for a mix of reasons. For instance, trees may have reached the ideal age and size to be harvested but may be inaccessible due to poor or non-existent road access or there are roads but the distances to the nearest sawmill are so great that it does not make economic sense to transport the timber to the mill. A 2017 study found that the average distance that forest logs were transported in Europe by road was 121 kms.[7] Transporting timber distances over 200–300 kms is only economically viable by train or ship and these options may well not exist. In some cases the local forest industry is simply not geared up to fell and process more trees. It is a moot point – and clearly not one that everyone would agree with – but should we consider allowing country X to harvest *above* the NAI if country Y is harvesting *below* their annual increment? Would one cancel out the other?

Country	**2000**	**2010**	**2020**
Czech Republic	56	63	104
Germany	53	61	96
Netherlands	37	39	94
Estonia	62	52	85
Belgium	58	63	75
Hungary	66	57	75
Latvia	63	57	75
Austria	41	72	74
Portugal	66	61	71
Sweden	58	65	68
Poland	66	47	64
Slovakia	44	69	64
EU-27	51	53	63
UK	51	63	62
Croatia	36	41	60

Finland	60	54	58
Slovenia	27	35	56
Denmark	95	53	54
Luxembourg	36	78	49
Ireland	52	37	46
Bulgaria	30	43	45
France	60	50	45
Spain	36	37	41
Greece	54	30	36
Lithuania	40	52	36
Romania	38	34	35
Italy	27	25	34
Cyprus	9	5	5
Malta	no figures	no figures	no figures

Table 7:1 Percentage of the net annual increment harvested in EU forests plus UK

Source: Eurostat & Forest Research.[8]

Sweden and Finland are often criticized for how much timber they fell but both harvest less than 80 per cent of their NAI. Felling rates over 90 per cent often cause an outcry from some but it means the country concerned is close to optimizing their ability to produce sustainable timber. (Would you criticize a farmer for optimizing sustainable food production on their farm?) Conversely, countries with a *low* use of their annual increment could (should?) be criticized for not deploying their NAI to help optimize sustainable timber production to provide the key material needed to decarbonize our built environment.

Overall, we can observe that while the figures may vary, they all point to the felling rates being below the 100 per cent point at which harvesting would be exceeding growth. So we can say that the current level of timber extraction from Europe's forests is sustainable. In addition, we need regularly to remind ourselves that our forests are not just for producing commercial timber but are also ecosystems that provide valuable habitat in terms of biodiversity hence extraction levels are not the *only* sustainability criteria, as noted earlier, that Europe's forests should be meeting.

As part of this debate, we also have to confront the peculiar and counterintuitive situation that while Europe's forests are expanding in size their ability to store carbon is declining. By 2025 some calculations suggest that European forests will hold 18 per cent less carbon than they did in the early 2000s.[9] How can this be possible? The primary reason is that we now have more older trees whose ability to absorb carbon is lower than it was when they were younger. Our forests are at carbon saturation point. This is then exacerbated by pests, disease, fires and storms – the four horsemen of the forest apocalypse – all of them on the increase due to climate change. They arrive, wreak havoc, then depart leaving behind them chaos. Some of this timber can be salvaged and put to good use, but the majority of it will be at best underused and at worst left to rot on the forest floor releasing methane into the atmosphere (Chapter 2). And the four horsemen are indiscriminate in their onslaught, removing just as many trees in their prime for sequestrating carbon dioxide. Getting Europe's forests back on course to sequester a rising amount of CO_2 is a tricky task involving some wise SFM. But it is not impossible and, in part, can be delivered by expanding the forest cover with new additional growth which will, because it is young, absorb a disproportionate amount of CO_2. At the same time the existing forest, from which we extract timber, has to be managed in a way that balances nature conservation (biodiversity), climate challenges and the growing demand for wood all in a sustainable, circular way.

Having examined the current level of timber extraction from the European forest we can see that further timber could be extracted without passing the continent's annual increment and therefore some additional supply of sustainable timber is available. But will it be enough going forward? To answer this question, we need to have a better understanding of the level of demand that is coming.

Writing in the *Investment Monitor* Mitchell states:

> The World Bank forecasts that global timber demand will quadruple by 2050, driven by economic and population growth. Gresham House, the London-based specialist alternative asset manager, has a lower estimate but is still forecasting that global timber consumption will surge by 170% over the next 30 years, driven by urbanization, decarbonization and housing demand. It forecasts timber consumption will increase by 3.1% a year over the next 30 years up to 2050, up from 1.1% a year during the past 20 years.[10]

This demand is coming from the "bioeconomy". A broad term, but essentially the bioeconomy seeks to replace products currently made with fossil fuels with renewable nature-based materials instead, materials such as wood.[11] The search is therefore on for ways in which we can replace – substitute – petroleum in the manufacture of plastics, detergents, paints, cosmetics, food packaging, clothing, fuels and pharmaceuticals.[12] Also included are the materials for use in the built environment that we have been discussing: more timber frames, more engineered timber, more wood fibre insulation. For many of the applications in the bioeconomy listed above we should note that the wood needed does not have to be in the form of a solid plank or log (as is needed for construction timbers) but can be supplied as wood fibre like that used in paper manufacture. This is an important point in the context of sustainable wood supply as we shall see later.

Timber-framed and mass-timber houses, insulated with wood fibre insulation, make up a sizeable component of the bioeconomy. Indeed, they were used as the first example of a bio-based product in the European Commission's Updated Bioeconomy Strategy in 2018:

> A sustainable European bioeconomy is necessary to build a carbon neutral future in line with the climate objectives of the Paris Agreement. For instance, in the construction sector engineered wood offers great environmental benefits as well as excellent economic opportunities. Studies show that the average impact of building with 1 ton of wood instead of 1 ton of concrete could lead to an average reduction of 2.1 tons of carbon dioxide emissions over the complete life cycle of the product (including use and disposal).[13]

The potential for the bioeconomy globally is staggering. A 2020 McKinsey Global Institute report estimated that by 2030–40 the global bioeconomy could create approximately $2–4 trillion in direct annual economic impacts.[14] This bioeconomy is underpinned by two key sources of biomass, that of the farm and the forest. In Europe 70 per cent of the bioeconomy's biomass is grown on farms (via crop production, not wood) and 27 per cent in the forest (wood).[15] The bioeconomy's demand for sustainable wood is only going to increase (which in many ways is a good thing) however the myriad of opportunities that it opens up will call for larger and larger amounts of woody biomass as one of its key feedstocks (raw materials) which brings us back to sustainable wood supply – will there be enough?

It is my view that harvesting more timber from Europe's existing forests while remaining within the annual increment *will not* provide enough sustainable timber for the future demands of the bioeconomy, including a significant increase in timber for the built environment. Where then might it come from? Before February 2022, part of the answer to that question would have been "Russia", which has one-fifth of the world's forest, more than the Amazon, and is the largest forested region on earth.[16]

Historically Russia has provided timber for Europe but European sanctions that followed Putin's illegal invasion of Ukraine have closed off this line of supply, including from Belarus, Russia's surrogate. Prior to the invasion 9 per cent of Europe's sawn softwood, mainly used in construction, came from east of a line stretching from the Finland/Russia border in the north to Kyiv in the south. At the time of writing, as the sanctions work their way through the systems of trade, we may see Russia switching their sawn softwood exports from Europe to China but the transport routes to achieve this are poor. Ukraine also has large forest reserves but these are essentially off limits due to the war. In addition, the EU requires that the timber it imports to have been sustainably produced. Russia and Belarus have struggled on this issue in recent years. Ending the war and solving these issues, including achieving PEFC status, could open up considerable additional supplies of sustainable wood.

However, the greater impact of Putin's war on Ukraine has been the decision taken across European countries to jettison their reliance on Russian fossil fuels, which pays for Putin's war. In the short term this has led to an increase in burning wood for energy as European countries look to find alternative sources of fuel while simultaneously seeking to drive down overall demand for fossil fuels. In theory Putin's war on Ukraine should speed up the UK's poor rate of home insulation and the EU's extremely slow Renovation Wave, which has at its heart the better insulation of 75 per cent of the European housing stock (see Chapter 3).

As heating bills have risen in response to huge hikes in gas and oil prices in the wake of the conflict, more European homeowners are buying wood-burning stoves and re-opening closed-off fireplaces to burn wood for heating, which has raised concerns about the impact on air quality of burning wood in domestic settings.[17] Much of what is burnt comes in the form of pellets made from "waste" woodchip, sawdust, crop straw and so on. Where wood and wood pellets are burnt, this is better done inside a stove than in an open fire, and the newer the stove the better, as it will be more energy-efficient and have lower smoke emissions.[18] Wood is also burnt to create heat for industrial processes and to generate electricity (see Chapter 8). In the UK there are laws and regulations which limit emissions into the atmosphere from such processes.[19] At a European level new standards setting limits on emissions were introduced in 2021 under the Industrial Emissions Directive.[20] Biomass combusting industrial facilities use technologies such as electrostatic precipitators, venturi or cyclone separators or fabric filters to control particulate emissions within licence limits. No similar technologies exist for domestic wood-burning stoves where efficiency is highly dependent on the stove and quality of the fuel used.

As an aside, Chinese involvement in the European hardwood timber markets has been a growing cause of concern in recent years as their buyers often outbid their European counterparts.[21] This willingness to pay above the market rate for hardwoods has led many to suspect that the Chinese authorities are willing to subsidize these purchases, leading to French and Belgian oak logs especially being packed into containers and shipped to China. At any given time, half the timber being shipped on the high seas is on route to China. This phenomenal demand is in part down to the fact that China's own forests are off-limits for timber production. This is because China is prioritizing the re-establishment of its forest after many years of over-harvesting and neglect and prioritizing its role in absorbing carbon and offsetting China's own massive carbon emissions.

Even if we see a return and expansion of a sustainable supply of timber from Russia and Belarus – and from Ukraine – which in an uncertain world is far from guaranteed, Europe would still require other sources such is the likely demand from the bioeconomy to develop new bio-based solutions made of wood fibres.[22] Before we discuss the other ways in which we might increase our sustainable wood supply, we are duty bound to challenge ourselves as to how well we are using the *existing* supply of timber. How do we best utilize this precious raw material so that the entire wood supply chain can expand to supply the bioeconomy alongside the development of healthy and resilient forests?

For many the best use is the most efficient use. How can we use our sustainable but limited wood supply as efficiently as possible in the battle against climate change? Here we soon realize that there is no "one size fits all" answer. Efficient use is intrinsically tied to the local, regional or national market in which the timber is being grown and sold – and the wood products manufactured and sold. Likewise, this year's efficient use may be different from three years ago and different again in three years' time. For instance, if the economy is strong then the house building sector is likely to be buoyant and there will be a strong demand and good prices for timber sold for use in the construction of family homes. However, if the economy has stalled, the winter is cold – at a time when Russian oil and gas is not available (hence energy prices have risen) – then selling timber for energy production may deliver the best financial return. Similarly, an economic revival, combined with building regulations being amended to make building wooden mid-rise apartments easier, could trigger a greater demand for engineered timber hence the best timber prices might be achieved if the wood is sold for CLT manufacture. Conversely if the bark beetle ravages forests this will necessitate premature felling with the wood potentially being best chipped and used to make wooden panels.

Most discussions concerning the efficient use of wood need to address the difficult and sometimes fraught issue of burning wood for energy. Here the key question is whether, when using wood for energy, we are using it as efficiently as possible in order to reduce our dependency on fossil fuels. Again, we can only answer this question within the specific geographical framework of a given market at a given time. Burning wood for energy is considered a "green" alternative to fossil fuels by the EU and governments across Europe, including in the UK. Burning wood, the same as burning coal or gas or oil, *does* release CO_2 into the atmosphere, but because the wood that is burned can be replaced relatively quickly with new growing trees which then sequester CO_2 from the atmosphere, this is seen to produce a "neutral" emissions cycle. You can burn forest wood and then plant new trees so that you have replacement wood ready in 25–50 years but you cannot burn a lump of coal and then make a new one in less than 50,000 years – likewise with oil and gas.

Currently Europe harvests some 562 million m3 of sustainable timber each year – well within the NAI for the continent as a whole, i.e. 60–65 per cent of our sustainable wood supply. Anyone harvesting trees understandably wants as high a sale price as possible for the felled tree and the greater the percentage of the tree that can be sold for use as timber – where the primary use will be in construction – then the higher the sale price. Roughly speaking, the longer the life of the product made from the timber, the more income can be generated; the shorter the life cycle, the less income. Hence there is a well-established hierarchy of use as the seller seeks to maximize their income and the buyer seeks to maximize the amount of timber that they have bought that is then turned into a saleable product. This leads to an inbuilt *efficiency of use* dynamic as the forest-based industries set about maximizing the product production from every tree they have purchased and hence maximizing their income from every tree processed (see Figure 7.1).

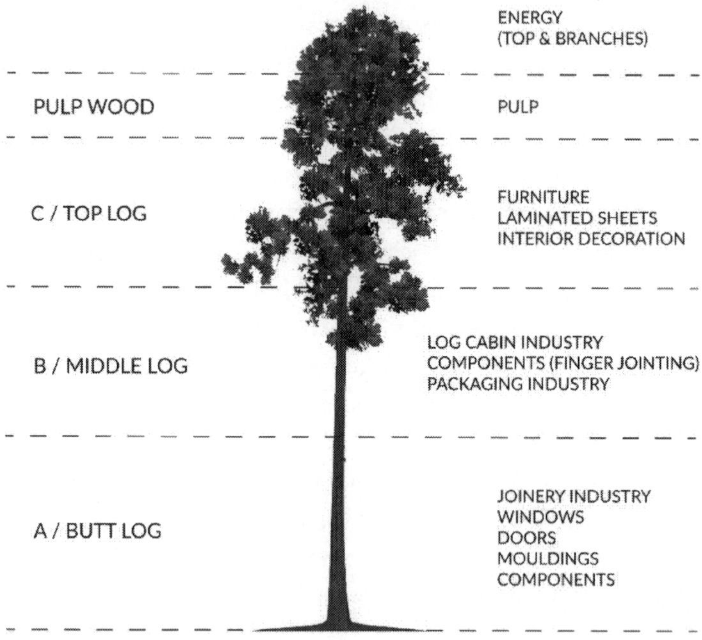

Figure 7.1 How a tree is put to efficient use

Source: Wood From Finland.

There is a parallel here, although maybe not to everyone's taste, with the meat industry. Having visited an abattoir I can vouch for the efficiency of the operation. Nothing goes to waste, everything is sold on to a range of buyers for a mix of uses but the priority is always to maximize the amount of quality cuts of meat – the prime cuts – as this is where the most money is to be made for the farmer and the abattoir. It is the same with a tree, with the timber for use in the built environment being the "prime cut".

Across Europe more than 70 per cent of our timber leaves the forest and enters a production process to manufacture a wood-based product: timber for construction, cardboard for packaging, fibre for hygiene products – including toilet paper and sanitary towels – paper for books, newspapers and magazines, etc. Many of these products will be substituting for fossil fuel-based carbon-intensive alternatives. In the battle against climate change this is without doubt an efficient use of our sustainable wood supply and the more we can improve recycling and re-use rates of wood-based products the more circular and efficient this process will become. That said, using less of all materials should be our collective starting point, i.e. material efficiency.

At the same time less than 30 per cent of timber removed from the forest is burned directly. The majority of this energy wood is delivered to people's homes where it is used as fuel for heating and cooking. This has been the tradition for hundreds of years and it makes good sense in rural areas, including in Scandinavia and Europe's eastern and Alpine regions. Most of this timber would not be suitable for making products, i.e. not big enough, not straight enough, not healthy enough. It is low quality wood, of relatively low financial value and can include treetops, branches and thinnings. Relatively little – about a quarter of the 30 per cent – goes directly for use in power stations.[23]

In the factories producing wood products we also have what are known as "residues" which are the by-product of the manufacturing process (the bits left over after the wood is cut to make the product). These residues, also referred to as "side-streams" or inaccurately as "waste", are often burned on site to produce heat for kiln drying the timber as it arrives from the forest before it is processed and/or used to produce electricity on site to power the manufacturing process. Any surplus energy can be sold into the wider electricity grid, or the residues can make their way off-site to a power station that uses wood as a fuel. Either way these residues are very efficiently used with 95 per cent of the energy content of the wood residues being utilized. The whole system is highly efficient as it always chases the higher value sales, which are nearly always achieved when the timber is sold to make products, rather than when sales for energy production in power plants. Yes, some forest timber that could go into products does get burned for energy in power stations but, as noted above, this is a relatively small percentage of the whole and usually only happens in one of two scenarios: where the wood supply is so high that all the other demands locally (demands that pay a higher price) have been met and within the geography concerned there is a surplus of timber; or it can occur when the product markets for residues are a long distance away from the place where the residues are created. In these cases the financial cost and the environmental impact of transporting low-value residues hundreds of kilometres makes no sense and hence burning them to generate energy in situ is the most efficient thing to do. Note that in the UK very little of our home-grown forest timber is burned in power stations to produce electricity.

With both Europe and North America facing warmer summers and an increased danger of forest fires we will increasingly need to remove thinnings from the forest rather than leaving them on the forest floor where they increase the risk of fire.[24] These thinnings are the product of SFM and are usually generated several years after the trees are planted and several years *before* they are harvested. The main objective of thinning is to reduce the density of trees and thereby improve the quality and growth of the remaining trees to produce a saleable product.[25] In addition, when the final felling takes place it will in future be necessary to remove more of the resulting residues, i.e. a forest tree harvesting machine will usually strip the branches from a conifer and these branches are left on the forest floor, along with the top of the tree, while only the central tree trunk is actually removed from the forest for processing. These residues are ideal for energy production and their use will enable we to use our sustainable wood supply more efficiently. There will, however, be an environmental loss resulting from less dead wood on the forest floor, which over time rots down and feeds the soil. But, as noted, in the short term it is a fire risk and in the longer term as it rots down it releases the GHG methane into the atmosphere. It should be noted that the environmental impact of a devastating forest fire is far greater than the environmental impact of increasing the amount of thinnings that are removed following both SFM and felling. Research righty continues in this area weighing up the advantages and disadvantages of removing residues from the forest floor. Recently Węgiel *et al* found that in the Polish forest "the increased removal of biomass from the forest environment does not necessarily result in its rapid degradation".[26]

Undoubtedly Europe can increase the efficiency with which it uses its existing supply of sustainable wood and we should continue to challenge ourselves on this in future, for example, by switching to the more materially efficient LVL rather than CLT where feasible (see Chapter 4). In addition, burning wood for energy is *not* intrinsically a bad thing to do. However, burning wood for energy inefficiently *is* bad practice. Ideally – and this may seem contradictory to what has been said before – *all* harvested wood should *eventually* be used for energy. The climate "trick" is to delay the length of time between harvest and burning for as long as absolutely possible. Ideally a tree harvested in 2024, planted in 1984, will become either a timber frame in a house or engineered timber in a block of flats for 50, 75, 100 years. Then if the life of the building can no longer be extended through renovation the timber should be removed and in the best scenario *reused* as a timber product in a new building. For instance, a wooden floor could be removed after 80 years and reused by being relaid as a floor in a new house, where it may exist for another 50 years. When it could then be removed and this time *recycled* into a wooden panel as part of, say, a kitchen, this time for 20 years before *again* being recycled, maybe *twice*, before finally the wood fibres break down to such an extent that it can no longer be recycled as a product. But what a life! Two hundred and thirty years storing carbon and 190 years as a range of products substituting for carbon-intensive alternatives before eventually being burnt for energy. This is more efficient than going to landfill as waste where it will rot and potentially release methane. Instead wood being recovered for energy use at the "end of life" should be seen as a major future source of energy.

Clearly, we need to become much better at *reusing* and *recycling* our wood-based products – an approach which can contribute to what is increasingly referred to as "urban mining".[27] Rightly there is growing interest in what happens to timber used in construction at its "end of life". There are typically five main options, in the following order of preference: prevention (i.e. retain the building and do not demolish it), reuse, recycle, recovery and, finally, disposal.[28] These categories do not only apply to timber. Environmental considerations, led by climate change, are challenging us to reconsider what happens to all materials used in the built environment at the end of life. Let's consider each option in turn.

Prevention is increasingly seen as key to our climate goals with demolition becoming the action of last resort not the default stating point. In 2023 the City of London Corporation, which aims to be net zero by 2040, instigated new guidance for demolition requesting developers' proposed schemes "show the impact of different development options with respect to carbon emissions" and "carry out detailed studies before submitting schemes for planning . . . and consider alternatives to demolition".[29] Stripping back buildings to their original structural skeleton before refurbishing them will become an increasingly common sight in our cities.

Reuse involves removing the timber components – structural or non-loadbearing – and reusing them for the same purpose in another building which itself may be either new or refurbished. Here the main issue is ensuring that if they are loadbearing structural components their mechanical properties can be guaranteed when they are redeployed. A parquet wooden floor from a Victorian school is easier to reuse as no structural testing will be necessary.

Recycling wood is typically the process of making woodchips, which can then be mixed with virgin wood to produce chipboard, technical name panel board, but also known as particleboard or low-density fibreboard, from which furniture can be made.[30] Every house in Europe and North America in all likelihood contains an item made from chipboard, fitted kitchens in particular. Chipboard has the added advantaged that it too can be recycled over and over again into more chipboard. In the UK 1 million tonnes of recycled wood, almost a quarter of the total, is used each year in this way, i.e. the recycled wood has itself already been recycled at least once.[31] Other recycled uses of wood include animal beddings, equestrian and landscaping surfaces, play areas and filter beds.

Italy is a European leader in wood recycling. Its national consortium Rilegno, formed in 1997, is tasked with the collecting, recovering and recycling of wood, including from the construction industry, industrial packaging and pallets, shipping companies, unwanted furniture, fruit and wine crates.[32] In 2022 it recycled 1.7 million tonnes of wood which equates to almost half of the UK's annual waste wood.

Recovery involves burning the wood as biomass to produce heat usually to generate electricity. To date this remains the primary use of the UK's waste wood, including timber removed from a demolition site. Waste wood for biomass accounts for approximately 70 per cent of the UK's waste wood use, equating to over 2.7 million tonnes of wood every year.[33] This market has seen significant growth since the first large-scale biomass plant for waste wood opened over 16 years ago. UK plants are compliant with the Industrial Emissions Directive and by using waste wood as their fuel they help to prevent the methane emissions that would result if the wood went to landfill. The energy produced is enough to power 840,000 households and accounts for 1 per cent of annual UK power consumption.

The final option disposal is little used. In the UK, in part as a consequence of increased charges for disposal to landfill, very little wood ends up being buried. According to the Wood Recyclers' Association, "Data from the Environment Agency's Waste Interrogator 2018 suggests that less than 1 per cent of waste wood ends up in landfill".[34]

The more we can avoid demolition in the first place by refurbishing existing buildings and the more we reuse and recycle wood then the longer the carbon in the wood will go on being stored. This will hopefully be incentivized by a more consistent representation in widely used life cycle assessments (LCA),[35] which assess the environmental footprint of a product from material extraction to waste treatment, such as environmental product declarations (EPDs) (see Chapter 3). Currently EPDs do not show a benefit for storing carbon in products such as wood, which needs to change. By using what is known as "dynamic" LCA, the length of time biogenic carbon storage is maintained – either within the product life or when re-used/recycled – is shown as a reduction in carbon footprint in the EPD. These more accurate results would show less atmospheric CO_2 due to the larger stored carbon pools created by such products and their extended use would be recognized, rightly, as a benefit.

This is an important issue because if we fail to fully and adequately recognize that wood products can be repeatedly reused and recycled we undervalue the contribution that timber can make to addressing climate breakdown, not only through the ability to substitute for more carbon-intensive alternatives, but through its ability to store carbon in wood product after wood product decades after the demolition of the original building that contained the timber, and up to 100 years or more since it was a tree in the forest. If we fail to value this contribution we should not then be surprised if we subsequently fail to maximize the opportunities for the reuse and recycling of wood.

With regards to the built environment the challenge for architects is to design wooden buildings today that in future can be disassembled and their component parts reused in a new or renovated building where these parts perform similar tasks. As dRMM architects see it, "We see opportunity for the end-of-life picture to be improved in the future, with more timber assets being repurposed at their highest asset value as part of a circular economy, i.e., mass timber structural panels able to be repurposed again as structural panels, rather than broken down to layers or as chippings".[36] This was certainly the thinking behind the Black and White Building in Shoreditch, London, winner of the sustainability category in the 2023 UK Wood Awards, in part because as the judges noted, "Both the superstructure and façade are also designed for disassembly, able to be deconstructed and redeployed in other projects as a kit of parts to extend their lifecycle".[37] In a similar vein sustainable design guru Lloyd Alter noted, "It's designed for disassembly, 'screwed not glued,' which means it can be taken apart and the components reused. This is critically important if the building is going to be truly considered sustainable".[38]

So that wood in buildings can be repurposed more easily there are some obvious considerations, for example avoiding the use of metal nails to fix wooden frames, boarding, etc. Although a tried and tested, quick and cheap process, a nail is much harder to remove than a screw when you come to disassemble. And a nail in a piece of wood that you want to chip to make chipboard is an unwanted impurity that is difficult to remove, whereas if the original construction had used a wooden dowel then it too can be chipped and recycled into a panel board.

The single biggest climate change gain we can make from the bioeconomy is by optimizing the percentage of timber that is used to make long-life harvested wood products for the built environment. Here, the products will both store carbon and substitute for carbon-intensive alternatives, such as concrete and steel. Currently around 15 per cent of timber makes its way into long-life harvested wood products that in turn last for 50, 75, 100 years. As the European Confederation of Woodworking Industries (CEI-Bois) has noted, there is significant potential to increase the amount of wood used in both new build and renovation. Currently, 0.5Gt of CO_2 equivalent is taken out of the European forest each year in the form of sustainable timber. CEI-Bois argues that we should be aiming to at least double this figure by 2050 so that wood can play its full role in tackling climate breakdown by making a significant contribution to decarbonizing the built environment.[39]

Currently the majority of Europe's forest timber is used in short- or medium-life products, for example, cardboard packaging for all those online shopping deliveries, or is burned for energy. This state of affairs is used consistently as a line of criticism by *some* environmental NGOs.[40] Rather than argue *for* an increase in the amount of timber used to make long-life products they argue *against* large swathes of the products that compose the rest. Admittedly if we did burn 85 per cent of our forest wood and used the other 15 per cent only for short-lived products with no recycling then these green NGOs would have a case for "keeping the trees in the forest" but this is *not* the situation. Rather we need the environmental NGOs to get behind a campaign to increase both the percentage and the absolute amount of sustainable timber being used to make long-life wooden products where carbon can be safely stored for decades while simultaneously replacing alternative materials that have a high level of embodied carbon.

What this must not become is a debate about who gets less wood, which it may do if we fail to grow the supply of sustainable wood to meet the expanding demand. By increasing the supply of sustainable wood we can increase both the amount and the percentage making its way into long-life harvested wood products *without* reducing the amount available for short- and medium-life products (although the share used for the latter could go down).

To conclude, Europe at present has enough sustainable forest wood to meet the current level of demand. However, this situation is changing as the demand from the bioeconomy for sustainable wood and wood fibre is rising. Within this wider demand growth also sits the growing need for more wood to optimize the role of timber in decarbonizing the built environment. As a result, we are going to need significantly more sustainable wood and must grow the supply. In part we can meet this demand by:

1. Managing our forests according to the standards set by Sustainable Forest Management (SFM) and investing in making our forests more resilient.
2. Increasing the percentage of the net annual increment (NAI) that we use but not by surpassing it.
3. Returning and establishing a sustainable wood supply from Russia and Belarus – and separately from Ukraine – but only after the war in Ukraine ends, Russia withdraws and peace is restored.

4. Optimizing the efficient use of our existing supply of sustainable timber.
5. Increasing alternatives to existing "end-of-life" use.

Taken together these five initiatives will increase the amount of sustainable timber available, but it will not be enough. In the chapters that follow, we shall explore additional ways of increasing supply.

At the same time as increasing supply we need to see the percentage of timber being used to make long-life harvested wood products – compared to other products – increase. In this way we can ensure we are delivering the maximum climate benefit by moving more and more carbon into storage in the built environment. It is these construction and renovation products, including insulation – not the medium- or short-term products – that can both store carbon for 50–100 years and substitute for more carbon-intensive alternatives, especially concrete, steel, plasterboard, brick and block. This is how we will best optimize the use of our sustainable timber in the context of climate change. Given that we know there will also be a rise in demand for wood for energy (in part due to bioenergy with carbon capture and storage, see Chapter 8, and in part from the continued shift away from coal) we also need to ensure we can meet this specific demand. Currently our wood for energy grows predominately in the forest. As we seek to ensure its increased future supply, we must consider other potential sources. In Chapter 8 we shall explore a specific way in which we can do this within the next ten years.

8

Increasing wood supply

In Chapter 7 we identified that the expanding bioeconomy, which includes using more timber to decarbonize the built environment, is going to demand larger and larger amounts of sustainable wood and wood fibre in the years ahead. We concluded that additional sources of supply, over and above those identified in Chapter 7, would need to be found. In this chapter we examine one specific way in which we can deliver a significant increase in sustainable wood supply: by growing more wood on farms.

We first need to remind ourselves that wood comes in all shapes and sizes, or to be more precise, the demand is for different sorts of wood for different uses. To build more timber-framed houses and to erect more big buildings made from engineered timber we will need an increase in the supply of traditional timber, i.e. long strips of cut wood. In future, as today, the vast majority of this kind of wood will be sourced from the forest. We will also need more woodchips to make wood panels and wood fibre to make insulation. Both of these products *can* be made from tall "straight" trees grown in the forest but they *could* be made from smaller trees that are not as straight grown on farms. Similarly, wood for energy *can* be sourced from the forest but *could* also be sourced from the farm. In the same way, the wider bioeconomy *can* take its wood fibre from the forest but *could* – in most cases – take if from the farm.

Let us begin by focusing specifically on meeting the growing demand for wood for energy, often referred to as *biomass*. Biomass is simply organic matter used as a fuel and it can be supplied by crops such as willow, Eucalyptus, Miscanthus (a large, woody plant from the grass family), hemp, straw and maize. As we have seen in the previous chapter this demand is set to grow. As this wood for energy demand increases we need to ensure that there are no future conflicts of demand. We must avoid competition over the supply of wood between wood for the energy sector on the one hand and wood for the construction sector on the other. We must ensure that going forward there is enough supply for both. According to Bioenergy Europe in 2018 70 per cent of our biomass-generated energy came from the forest, 18 per cent from the farm and 12 per cent from waste materials.[1] If we increase both the percentage and, more importantly, the overall amount that is sourced from the farm we could in a mere ten years provide the primary source of biomass needed to meet the expanding market for wood energy across Europe from the farm, which we could not do from Europe's forests.

Farm-grown biomass *crops*, such as hemp, straw and maize, can be grown on an annual cycle. Farmed *trees* require longer cycles: three years in the case of willow; five to seven years for poplar. This is often referred to as "short rotation coppicing" (SRC) for the willow cycle or "short-rotation forestry" (SRF) for the poplar cycle. All such bioenergy farm crops would normally have faster growing times than forest wood, which in Europe is usually grown on 30–100-year cycles, although the southern United States has growing cycles of 25 years. That farm biomass can be produced more quickly is in part due to farms having established themselves on the better soils, leaving forests to develop on the poorer soils.

As we look ahead to where our biomass for energy may come from in the future we have some choices to make. Table 8.1 forecasts a 14 per cent increase in overall biomass use by 2050 and posits two possible scenarios for supplying that demand. In the first forest biomass drops from 71 per cent to 3 per cent and farm biomass increases from 16 per cent to a staggering 73 per cent. This is the direction of travel in which we need to move. Scenario two, shown in the column on the right, shows a 66 per cent increase in *forest* wood for energy, which is not something we should be aiming for, not least because that amount of forest wood is simply not available unless existing users such as the wood industry were to use considerably less wood. Some will question the need for the massive increase in farm-produced biomass set out in the right-hand column but for those of us who have grasped the size of the coming bioeconomy demand for wood fibre, including for biomass, it does support the notion that a big increase in wood fibre grown on farms is both realistic and doable.

Type of biomass	2021	2050 Scenario 1	2050 Scenario 2
Forest biomass	105	5	174
Agricultural biomass	23	124	444
Waste biomass	20	40	119
Totals	148Mtoe	169Mtoe	737Mtoe

Table 8.1 Gross inland energy consumption of biomass in 2021 and potential in 2050 for the EU-27, millions of tonnes of oil equivalent (Mtoe)

Source: Bioenergy Europe Statistical Report 2023.[2]

Willow is the standout candidate if you want to grow woody biomass quickly in a farm setting. You sink a 30 cm long cutting into the soil with three centimetres showing above the surface and then – apart from holding back the weeds and pests for the first six months – you do not need to do much else. Three years later you have a bushy tree around five metres high with a central trunk that has a girth equivalent to a 20-year-old Sitka spruce grown in a forest. This impressive ability to put on size and weight at high speed has made willow one of the most popular options for farmers wishing to grow an energy crop. After three years you harvest it with a machine much like a combine harvester and then you chip it, dry it, and it is ready to burn for energy or other uses. One hectare of willow can provide the equivalent energy of 5,700 litres of fuel oil (enough to heat three houses for a year), with less of the CO_2 emissions associated with burning fossil fuels. The bonus from a farmer's or landowner's point of view is that you do not need to plant again to achieve a second crop as the "stalks" of the harvested willow will simply shoot away with new growth in the spring restarting the whole process. Once planted the willow basically looks after itself. To quote willow planter Neil Watkins of Eco Crops Ltd: "I hesitate to say, 'willow is a wonder crop' but when integrated into a farm's existing food production it can provide an additional income stream while at the same time delivering a crop that can replace forest wood which is increasingly in demand as a construction material".[3]

A perceived problem with willow is that its roots may interfere with a farmer's existing field drains. If the land in question is being permanently removed from food production – conventional crops or cattle – then this is not an issue. However, there are types of willow that do not deeply root such as *Salix viminalis* which spread their roots across the soil surface and do not interfere with drains.

There are barriers for farmers who are considering planting willow or a similar SRC, like poplar. At current prices it costs anywhere between £2,000–2,500 a hectare to establish such a crop. There is then a three-year wait before the first harvest and its income. This income will only result in the farmer breaking even and it will not be until the second harvest after six years before a profit will be made. While six years is a relatively short length of time to a forester, it is a relatively long time for a farmer. It is possible to increase yields with extra nutrition so a profit can be made after three years rather than six, but this nutrition, maybe manure, will need to be bought or taken from another use elsewhere on the farm.

Any mention of farming and costs inevitably raises questions around government subsidies including environmental schemes. Support payments to UK farmers are in a state of flux post-Brexit. Farmers in the EU remain within the Common Agricultural Policy. Both approaches – UK and EU – see farm support linked to farm practices. If we decide that growing willow on farms as a source of raw material for the bioeconomy, including for energy, is a good idea we can decide to incentivize it in such a way that farmers make money sooner, after three years and not have to wait six.

Whether we use willow or other energy crops the quickest way to grow significantly more biomass will be on our farms rather than in our forests.

This is starting to happen, as we can see at Drax power station in North Yorkshire in northern England (see Figure 8.1). Drax produces some 6 per cent of the UK's electricity. From its opening in 1974, it burned locally deep-mined coal. Now Drax burns forest-grown wood pellets – biomass – imported mainly from the USA. It is the largest biomass powerplant in the world and the largest single site producer of electricity in the UK.[4]

Figure 8.1 Drax power station in North Yorkshire, UK, with biomass storage tanks

Source: iStock; Teamjackson.

These pellets, like the coal products that preceded them, are ground into a dust which is then blown into a furnace and burnt to produce the heat to boil the water, turn the turbines and generate the electricity. In its coal days the power station had six turbines. Now it has four fired by wood.[5] Until recently, it had two on standby to burn coal if the national electricity grid was short of power. None of the wood pellets currently burnt by Drax originate in the UK. But in 2022 Drax began considering some "onshoring" – sourcing from the UK mainland – with the aim that a small amount (6 per cent) of the biomass burnt will come from the UK by 2027. This biomass will not come from UK forests but rather from UK farms within approximately 100km of the power station. Initially Drax has focused on Miscanthus as they burnt this for a brief period when they transitioned from coal, therefore some farmers in the area have experience of growing it. Drax have indicated they intend to source at least one further additional material for biomass which *could be* a short rotation tree "crop" such as willow or poplar.

The advantage of Drax shifting to 6 per cent farm material is that potentially this means 6 per cent less forest wood will be imported. But if the amount of wood, in real terms, burnt at Drax increases, the same amount of US pellets would still be needed. However if the actual amount being burnt remains constant then it will result in a real fall in the amount of forest wood from the USA being burnt. What would then happen to the unwanted 6 per cent is a moot point. Would the forest owner be able to find another buyer for their wood? What would this use be and what would the pros and cons be? Might it then be lost as forest altogether? Drax is also keen to encourage other biomass energy companies to switch some of their sourcing of fibre from the forest to the farm. In the case of Northumberland, in the north of England, this could draw in the Cramlington biomass power plant and the previously coal-fired Lynemouth biomass power station.

Drax's influence in the bioenergy world is substantial and where they tread first others will likely follow. In the UK this will put the majority of farmers within the catchment area of a biomass-burning energy plant, of which there are currently 78 in the UK. Presumably Drax are doing this because they have an eye on future government farming policy, including payments, which they will be better able benefit from as a collective of biomass energy producers.

The UK forestry world is somewhat sceptical about Drax's intentions. They worry that while Drax *say* they are only interested in using *farm-grown* UK biomass this might change in the future. What if they turned to *forest* biomass in addition to – or even instead of – the farmed equivalent? Everyone agrees we simply do not have enough of this in the UK. In fact Drax burns the equivalent of 127 per cent of the UK's entire annual wood production.[6] And if Drax was still in receipt of a government subsidy when shopping for biomass this would put them at a significant advantage over other existing customers of UK forestry. This is the kind of competition we need to avoid and the best way to avoid it is by increasing supply.

In the longer term Drax are hopeful that they will be among the first power stations in the world to perfect carbon capture and storage (CCS) at scale. This will enable biomass to be burnt and the carbon that is released to then be captured and pumped underground for safe long-term, some say permanent, storage. The flipside of this aspiration is that if CCS does not materialize as a viable option then Drax will almost certainly close. To date CCS has been tried and tested successfully in the laboratory and in larger industrial tests as at Mongstadt in Norway conducted by Statoil. Off the coast of Norway two CCS operations, at Sleipner and Snohvit, have been capturing a million tonnes of CO_2 annually for 27 years and 15 years respectively.[7] Without doubt we need CCS as part of the architecture that will deliver net zero, a view taken both by the UN and by the UK's Committee on Climate Change, but to date, CCS is progressing slowly and is expensive and unreliable.[8]

It should be noted, however, that Drax's current modus operandi arouses strong emotions, especially among the majority of environmental NGOs, which do not accept that this way of generating electricity, particularly its massive scale, can be classified as green or renewable. The scientific community is divided. While an MEP I well remember receiving a letter signed by 796 scientists from around the world opposing burning wood for energy,[9] but in 2022 there was an open letter from 550 scientists *supporting* burning wood for energy.[10] Part of what makes the whole Drax situation so complicated and emotive is that UK taxpayers, via the government, provide a financial subsidy to generate energy in this way. The UK is not alone in this; several European governments do likewise. Why? Because it was seen as a way of speeding up the move away from coal, the worst of the fossil fuels when it comes to carbon dioxide emissions. In this aim the subsidy has worked.

As each year goes by, however, one point seems increasingly illogical and raises important questions. Why subsidize burning forest wood for power and not subsidize making long-lasting wooden construction products that store carbon and substitute for materials with a high degree of embodied energy derived from the burning of fossil fuels? Where this observation seems glaringly anomalous is in Northumberland where whole virgin timber logs leave Kielder Forest on lorries. Some go to the Egger wood panel factory in Hexham and are made into kitchens that will store carbon, with no publicly funded subsidy. Other logs will go to the wood-burning power station at Cramlington where they are chipped and burnt for power, *with* a subsidy. In other words, it is not a level playing field. To be fair to Drax they very deliberately stayed away from this market where there was a strong domestic forest wood demand and went instead to the USA where the forest wood they source is not in demand by an Egger equivalent. Again, by increasing supply we can remove this competition that could in future create problems within the wood supply chain.

Government subsidies for burning wood for energy in the UK will be removed by 2027. This may explain, in part, why Drax are starting to consider sourcing some of their biomass from farms rather than forests. These farms will continue to receive subsidies or payments, in some shape or form, via the UK's replacement for the Common Agriculture Policy. The Climate Change Committee has also indicated there will be a growing demand for UK grown biomass: "Over the course of the Net Zero transition, our modelling suggests we need to increasingly rely on sustainable domestic biomass planting with carbon capture and storage to produce clean energy, while limiting biomass imports".[11]

At this point it should be noted that wood fibre grown on farms does not *only* have one use . . . to be burnt for energy. Just like its sibling from the forest, it can be used for many of the increasing and varied applications emerging in the wider bioeconomy.[12]

When it comes to quantity of material, an application where we will need vast amounts of wood fibre – way more than we currently have – is for wood fibre insulation for our buildings. In Chapter 3 we discussed insulation noting that when choosing what material to use there were two main considerations to be made from a climate perspective. First, how was the energy produced to make the product? Was it a green energy source or a fossil fuel source? Second, could the product store carbon? We concluded that in most cases a nature-based insulation such as wood fibre was preferable primarily because the alternatives could not store carbon. With the anticipated increase in future demand for insulation we will need to produce significantly more insulation. This will require a vast increase in wood fibre supply, more than our forests alone can provide and so it is to our farms that we need to turn.

A rapidly expanding insulation market based on woody biomass, led by companies such as Steico and Pavatax, could act as a catalyst for new factories in locations close to major markets, our big cities, and could free them from the need of being near major forests, as their wood fibre would be sourced from farms, which are nearer to cities than most of our forests. Given that the UK grows so little timber, and hence we have to import 80 per cent of what we use, it is hardly surprising that (as of 2024) we do not have *any* wood-based insulation manufacturers anywhere in the country. However, if they could access woody biomass produced by British *farmers*, rather than British foresters, a major barrier to entry would be removed.

The scale of this new market for wood fibre-based insulation takes a bit of grasping (see Figure 8.2). At a European level all countries are committed to improving the level of insulation within their buildings as part of their commitment to driving down fossil fuel use to hit the shared 2050 net zero target. Insulation, or the lack of it, rose up the political agenda in 2021, partly driven by campaigners from "Insulate Britain" blocking major roads that winter.[13] Subsequent soaring energy prices, in part due to Putin's invasion of Ukraine, have further focused minds on the need to insulate more, and faster. The EU have their ambitious "Renovation Wave" initiative, which seeks to renovate, including better insulating, three quarters of the EU's buildings – 190 million in total – by 2050.[14] Installing this insulation will drive down energy use, save households money and disproportionately help those on lower incomes for whom fuel bills are a significant and rising part of their annual expenditure. Lower fossil fuel energy use would, at the same time, mean lower carbon emissions.

In 2022 CEI-Bois did some work looking at the supply figures for the materials needed for this ambitious insulation push.[15] At an EU level the 190 million buildings that need to be better insulated could each require between two and five tonnes of wood fibre insulation. That works out at between 380–950 million tonnes of stored carbon. At the higher figure that is the equivalent of all the carbon currently stored in the forests of Sweden, France and Germany combined. To store this much carbon in the built environment – a massive CO_2 removal – would be a hugely beneficial climate action to undertake, but would there be enough wood fibre?

Figure 8.2 Wood fibre loft insulation being installed
Source: Steico.

The sustainable net annual increment (NAI) that the European forest currently yields is around 0.7Gt of wood but we already use 0.5Gt (see Chapter 7). This leaves 0.2Gt currently unused. But there are lots of other demands on this wood supply, not just the demand for wood fibre for insulation. Hence the vital importance of identifying other sources of wood fibre: the farm rather than the forest. The good news is that a significant increase in wood fibre grown on farms should provide us with enough biomass to meet not only the demand created by a huge increase in the manufacture of wood fibre insulation but also more biomass for energy. It should be reassuring for farmers that they are not reliant only on the biomass energy market. For instance, if the UK were to realize the entirety of its on and offshore wind potential plus a huge solar rollout then it's likely that we might not need to burn as much biomass for energy, but there would still be the wood fibre insulation market for farmers to sell into. This, in turn, would be part of the ever-expanding bioeconomy, needing wood fibre as a feedstock for many processes including textile manufacture.[16] In this scenario the farmer would have a choice of markets for their willow or other woody biomass. Wood fibre panel makers, like Egger in Hexham, could well become customers for farm-grown woodchips. Not *all* the virgin wood they use needs to come from the forest. It could also come from the farm.

Geography will in part dictate whether wood fibre insulation is made from material grown in the forest or on the farm. Both sources will be needed. Heavily forested countries are likely predominantly to use forest fibres whereas less forested countries, such as the UK, will need to use farm-grown wood fibre. Either way both approaches will sequester and store carbon and both approaches can contribute to the manufacture of long-life carbon storing products. Hybrid products, that bring together forest and farm fibres in the same item, will also no doubt be realized.

Although the demand for biomass as a raw material for numerous emerging uses in the bioeconomy will happen, and therefore be good news for farmers, biomass for energy may always be its primary use. This will almost certainly be the case if bioenergy with carbon capture and storage (BECCS) proves to be successful. BECCS raises the possibility of "negative emissions".[17] If you hold the view that burning woody biomass is green (i.e. when you burn it the carbon released does not need to be counted because you will plant new trees to absorb this emitted carbon) and *in addition* you capture the carbon when the wood is being burnt, then you have actually *reduced* the overall amount of carbon in the atmosphere. You have produced negative emissions. And if the likes of Drax receive one payment for generating electricity and *another* for capturing the carbon released when the wood is burnt, they could end up being paid (a great deal) to "remove" carbon permanently from the atmosphere, via BECCS.

Many environmental NGOs and others believe this concept to be deeply flawed.[18] This is partly because CCS has yet to get going at scale and at multiple locations and also because many worry that it detracts from the imperative of reducing greenhouse gas emissions now.[19] The difficulty for opponents of CCS and BECCS is that the UN have endorsed this approach as *one of* the many necessary approaches that will need to be deployed in order to reach net zero by 2050.[20]

To conclude, growing significant amounts of wood fibre on farms can provide the primary source of biomass to meet an expanding market for wood energy across Europe. This is not to say that some additional forest wood, over and above what currently is used for energy, cannot be used for wood energy. Some parts of forest-grown trees – thinnings, tops, branches – are ideal for this use (see Chapter 6). Whether we use forest wood or farm wood to meet most of this rising demand will be dictated by geography. Countries such as Finland and Sweden are so heavily forested that they may well have enough wood to meet the growing demands of all their markets – construction, renovation, insulation *and* energy – whereas countries with low forest cover like the UK, Belgium and Holland may wish to prioritize their forest wood for use in the built environment and use almost entirely farm wood for energy.

Having identified Europe's farms as the best place to grow the additional woody biomass we will need, first and foremost for energy wood but also for wood fibre for insulation and the other emerging bioeconomy demand, we find ourselves facing a new question: Do we have enough land to allow us to grow this additional biomass? This question we shall explore in Chapter 9.

9

Is there enough land?

Fact: *The amount of abandoned farmland in Europe, including European Russia, is equivalent to the existing farmland of France, the UK and Ireland combined.*[1]

In the last chapter, we explored how boosting the growth of wood fibre could provide us with the quantities required to significantly increase the use of wood in the built environment and the growing bioeconomy. This would involve wood fibre produced on European farms, in addition to that already grown in the forest. It would also be possible alongside the provision of biomass for the energy sector, until other energy sources could replace biomass from both the forest and the farm – if that is what we decide we want to do. It is likely that some percentage of energy will still be sourced from biomass all the way to 2050.

Do we have the spare land to produce the wood fibre we need? While the global population continues to rise, as does the population of the UK,[2] pressure on our finite land resources from competing demands is intense and where there is a high ratio of people to land, as in the UK, we need to be sure we are optimizing how we use our land in general. Wood fibre is only part of the picture. In this chapter, we will ask if there is sufficient land and whether it can produce enough wood fibre to meet the growing demand. In Chapter 10, we shall explore whether this can be done without negatively impacting food production.

Across the world there is a surprising amount of land that has been abandoned because human activity has ended. This includes areas which were used for farming, mining, manufacturing, settlements and so on: "since the 1950s, abandoned land has accumulated to up to 400 million ha [hectares] globally, an area roughly half the size of Australia".[3] At a European level, the majority of abandoned land is to be found in central and eastern Europe, including European Russia, and this is where the increase in wood fibre production could best come from. After the Second World War, these countries, all under communist regimes, nationalized their agricultural land, creating state or collective farms (only in Poland did a small number of private farms persist). This pattern of state ownership was not uniform and differed from country to country, but sowed the seeds for a fragmentation in land ownership once communism collapsed in the 1990s.[4] As state control gave way to the emergence of market economies, the countries of central and eastern Europe began a process of restoring property rights to former landowners. Whatever the ethical merits of such an approach, the practicalities were messy. Land formerly owned by one individual was now in the ownership of several children or many grandchildren, resulting in the break-up of many viable farm units. While the average size of a farm in the UK in 2005 was 56 hectares, it was only six hectares in Poland and Hungary, five hectares in Bulgaria and four in Romania where a mere 2 per cent of farms were larger than 10 hectares.[5] In 2017, Romania ranked top in the EU for the number of holdings, namely 3,630,000 farms, but bottom for the average financial output per holding at only €3,303.[6] On a visit to Romania in 2016, while serving on the Agriculture Committee of the European Parliament, I had the opportunity to see at first hand some of the former collective farms and how the country had moved on from state ownership. But it was also clear from that visit that a second wave of land reform – following the initial restoration of property rights in the 1990s – was needed to collect fragmented parcels of land and consolidate them into workable farms. Easier said than done, but this was achieved in Greece prior to their entry to the EU in 1981 and also in Estonia before they joined in 2004.

To make matters worse, the intervening years had in most cases seen the deaths or retirement of the former individual owners whose land had now passed to the next generation, many of whom had moved to the cities and had no experience of farming. The result was whole swathes of farmland being essentially abandoned – taken together the equivalent of all the agricultural land in the UK (19 million hectares), Ireland (4 million hectares) and France (27 million hectares). The single biggest area of abandoned farmland is to be found in European Russia at 32 million hectares, followed by Ukraine and Belarus, then Romania, Bulgaria, Lithuania and Latvia. According to Alcantara *et al.*, this land remains suitable for agriculture and so could be used for the production of wood fibre: "Our results showed that much farmland abandonment occurred in regions that are suitable for agriculture, highlighting the potential of Central and Eastern Europe to contribute to increasing agricultural production by recultivating suitable but currently idle farmlands".[7]

In addition, further west, Eurostat calculated in 2022 that 15 per cent of land in the EU was unused or abandoned with signs of previous use.[8] And a further 3 per cent of total agricultural land in the EU is projected to be abandoned by 2030.[9] Furthermore, a report from the European Commission in 2018 predicted that by 2030, 11 per cent of EU land classified as farmland in 2015 would be under "high" or "very high" threat of being abandoned. Seventy per cent of this land is currently arable. Interestingly, the predictions are that only 13 per cent of this land at risk of being abandoned will be converted into forests or natural areas.[10]

While these figures for abandoned land suggest substantial potential for wood fibre production, it would be a mistake to use all of the abandoned and underused land in this way. A large amount should be returned to what it was a century ago: forest. This in turn should comprise a mix of planting, split between commercial forestry and forest for nature (noting that biodiversity can be maximized in both). Former peatland should be restored, whereas other areas should simply be left for nature to take its course, which may or may not ultimately result in the return of some kind of landscape dominated by trees. Likewise, some recultivation would help increase food production, which is needed in Europe to reduce imports. It should be noted that abandoned farmland is not only a European phenomenon. The United States has lost almost 16 million hectares of farmland in the period from 1997 to 2018 – the equivalent of the landmass of England and Wales.[11]

So, across Europe, including in France and Spain, we have abandoned farmland on which wood fibre could be grown. Growing more wood fibre would undoubtedly increase the productivity of such land, which could simply involve planting willow or poplar trees. But what of existing farmland? Could that be made more productive? Can we increase the amount of wood fibre grown here? Posing this question raises the dreaded hydra of fertiliser, which is the obvious route to creating more growth on existing land. However, the good news is that wood could play a significant role in increasing productivity while also reducing the role of fertiliser on farmland.

Artificial or chemical fertilisers have been used extensively by farmers since the 1960s and have been both a blessing and a curse. They have boosted yields and allowed the ever-expanding global population to be fed. However, they have come at a high price, both monetarily and environmentally. The monetary cost has risen markedly in recent years due to the huge rise in the price of fossil fuels. It takes a lot of heat (energy) to make modern fertilisers. This cost is borne first by the farmer, but in due course is then passed on to the customer. The environmental costs are more complex but the climate cost alone is significant: "The global production of fertilisers is responsible for around 1.4 per cent of annual CO_2 emissions, and fertiliser use is a major contributor of non-CO_2 greenhouse gas emissions".[12] At 1.4 per cent it is greater than the annual CO_2 emissions of the UK and almost equivalent to the entire emissions of Germany. This carbon footprint comprises a combination of emissions during manufacture (fossil fuels are burnt to create the necessary heat) and when the fossil fuels are burnt to transport the fertilisers, often over long distances, to the customer on the farm. A further problem with modern fertilisers is that they create dependency. They are rarely a single, one-off application but quickly become an annual necessity with the farmer locked into yearly purchases. This is where wood comes in.

In the UK, one pioneer has attracted a lot of attention with a novel wood-based alternative to fertilisers. In the quest for answers to the fossil fuel-based fertiliser problem, researchers have been beating a path to meet Iain "Tolly" Tolhurst on the Hardwick Estate in the upper reaches of the River Thames. Two such pilgrims have been environmental author and activist George Monbiot and the Soil Association's Ben Raskin. Tolly has taken a piece of poor-quality land and transformed it into a successful horticulture business using only one external input: woodchip. The science behind this would *appear* to be that spreading woodchips in a thin layer on the surface of the soil, not digging them in, provides an interface between wood and soil which is ideal for the development of soil fungal populations. Their presence then boosts the earthworm population and increases the presence of helpful bacteria. Technically these woodchips are known as "ramial chipped wood" (RCW). I cautiously use the word "appear" because we are not sure yet of the science at work here. However, clearly something beneficial is happening. As Tolly told Monbiot, "The woodchip isn't fertiliser. It's an inoculant that stimulates microbes. The carbon in the wood encourages the bacteria and fungi that bring the soil back to life. We aren't feeding the crop, we're feeding the soil".[13] This woodchip could be formed from any type of wood, especially if it is a mix, but conifer alone is not recommended due to issues with polyphenols.[14] This is fine because we have other uses for it. Willow, however, is one of the better woods to use for RCW, therefore Tolly has started growing willow specifically for this purpose. As Raskin notes, "If you plant with the intention of producing RCW, single-species rows are the most practical and efficient approach. Willow seemed the most popular for its rapid growth rate and ease of chipping".[15]

Expanding this approach could give farmers much more control over their inputs and, crucially, costs. Rather than buying fertiliser, farmers could grow their own wood to chip. This would reduce the dependency culture created by the big fertiliser companies. Moreover, because farmers can produce their own RCW at a stroke, this home-grown approach would significantly reduce the CO_2 emissions generated by transporting fertiliser from factory to farm, all over Europe and beyond.[16] For Raskin, "there is still much to learn about using ramial chipped wood. There is no doubt it has massive potential to harness a material with little current value to significantly boost soil health, increase yields and help to mitigate climate change".[17] So, farms could grow wood fibre not only to replace forest biomass for energy production, but also as an enhancer of soil fertility.

We now have two routes to increase the availability of wood fibre. The first uses abandoned and underused land to grow wood fibre all across Europe but especially in eastern Europe. The second involves increasing the fertility of existing farmland soil by using wood chip which in turn can increase wood fibre yields wherever it is grown.

There is also a third route, but before we discuss it, let us pause and take a look at the crucial component of all this land: soil. All human life on earth is dependent on the 30 cm of soil that exists on the earth's crust. Destroy it and we all starve, rich and poor alike. We were warned by the experience of the American Dust Bowl in the 1930s that if we overuse the soil and fail to care for it, it can quite simply blow away and leave behind an arid wasteland. As rightly observed by Valujeva *et al.*, "Within the context of sustainable land management, soil is the most important resource that provides food, feed, fibre, water purification and regulation, nutrient cycling, carbon sequestration and regulation, and habitat for biodiversity".[18] Soil is a limited resource. Therefore, in addition to looking after it, we need to put it to the best use. We *know* how to look after it, but simply fail to do so. If we waste land, fail to make the best use of it, we will not be able to tackle climate change or biodiversity loss, nor will we manage to feed, house and clothe the world's growing population. There are multiple challenges at play here but the bottomline solution is that every square metre of land needs to be assessed in terms of what would be the best use for it. Often we can immediately scale up from a square metre to consider larger plots of land. However, by always reminding ourselves that even a square metre has a role to play in sustaining life on earth, we can ensure that everyone is playing their part.

Having focused so far on the bigger picture, it is worth noting that the issues in this book are also relevant at a micro level. For instance, is your garden, if you have one, playing its maximum role? Is the land owned by the nearby school playing its maximum role? Is the land owned by the local council playing its maximum role? Essentially, anyone with land, however small the area, needs to consider how they can best use it, for example, could trees be planted? It is worth noting that Tolly began by sourcing his wood chips not from farms but from people's gardens.[19] Just as we are used to rural farms providing urban food, we now need to consider the possibility of urban gardens providing "food" for rural farms in terms of woodchips to increase the fertility of farm soils. This would work particularly well for farms located on the urban fringe.

However, as we consider how best to put our soils and land to use, we run up against a serious obstacle to such innovative projects. The UK, along with many other European countries, does not have an overarching land-use strategy, a coherent framework within which to approach land-use questions[20] (although Scotland does[21]). Rather, most countries have agriculture strategies, which are separate from forest strategies, which are separate from biodiversity strategies, which are separate from house-building strategies, which are separate from climate strategies. Nothing is joined up; a silo mentality that too often characterizes government policy. What is clear is that it will be our rural areas that will provide the greatest contribution in terms of carbon sequestration, food production, timber growth and support for biodiversity.[22] All these different land uses, referred to as "multifunctionality", are not mutually exclusive, even though our siloed governance systems tend to make them more so. In addition, one should add recreational use, as the benefits of fresh air and exercise for physical health and mental well-being are well-known. Clearly, not every hectare should be or can be dedicated to any one of these primary land uses, rather we need a balance. Quite what this balance should be is a tricky question. The UK has a relatively large population in ratio to its land mass, being one of the most densely populated countries in Europe, along with the Netherlands and Belgium. Europe, as a whole, is less densely populated than the UK and some European countries have relatively small populations given the size of their land mass, for example Norway, Sweden, Finland and the Baltic States. Consequently, a one-size-fits-all land-use model is unlikely to work at a European level. Nonetheless, how we apportion land use is fundamental to how we tackle climate change. In his book *Regenesis: Feeding the World without Devouring the Planet*, George Monbiot writes: "I have come to see land use as the most important of all environmental questions. I now believe it is the issue that makes the greatest difference to whether terrestrial ecosystems and Earth systems survive or perish".[23]

This elevation of the land-use question by Monbiot will receive widespread support, even from many who, especially in the farming community, tend to oppose his views. Of course, the disagreement is then around how to use the land. For Valujeva *et al.* careful land management is the best way to mediate between competing demands regarding land use. Indeed, the EU's Green Deal requires that farmers address the competing demands of producing food, reducing GHG emissions and protecting biodiversity.[24] In the UK, for example, we produce around 58 per cent of the food we consume.[25] Should this be increased? It is worth adding that producing more food does not necessarily mean using more land. We can grow more food under glass and indoors using relatively intensive (but sustainable) methods which use relatively little land.[26]

So far, we have established that we need to boost wood fibre production by utilising more abandoned land and making *existing* farmland more fertile by using woodchip. We also need to do this within a coherent structure of land-use policies and a culture in which every square metre counts, including our back gardens.

The third main route towards producing the increased quantities of wood fibre we need is ultimately connected to land-use policy, but the driving force is closer to home, as it involves our diet and a big cultural shift: we need to eat less meat. It is a change we almost certainly have to make, and it has the support of the UK's independent Committee on Climate Change, who advise the government. By doing so, we would free up a significant amount of land for other uses. It has been calculated that if the global population were to eat the same amount of meat as the average American, we would actually need another planet. If Americans were to reduce their meat intake by just 25 per cent (and they are amongst the biggest meat eaters in the world) this would free-up over 9 million hectares of high-quality land, an area roughly the size of the state of Indiana or three times the size of Belgium.[27]

The United States is the home of cattle ranching, but the UK has its sheep farms, which have also come under scrutiny. Monbiot, again,

> My estimates suggest, conservatively, that some 4 million hectares of hill and mountain in the United Kingdom are used for sheep farming, or 22 per cent of the entire farmed area. This is roughly equivalent to all the land used to grow arable crops in this country. It is more than twice the size of the whole built environment and twenty-three times the area we use for growing fruit and vegetables. But in terms of calories, lamb and mutton, across the uplands and the lowlands, supply just 1 per cent of our food.[28]

This quote carries quite a punch and the extent of sheep grazing in the UK is likely to come under increased pressure from alternative land uses. Moving from 100 per cent sheep to 100 per cent something else, even trees, is almost certainly not the answer. Fewer sheep would release more land to plant commercial forests, for sure, and given that the UK is the third biggest importer of wood in the world, we need to be growing more of our own timber, and re-establishing more forest cover – including conifers – would help increase biodiversity levels. But, an increase in the number of trees (not necessarily evenly spaced but rather in clusters and strips in a deliberately integrated approach) on existing sheep grazing land would be sensible because, as it happens, trees and sheep are a rather good mix.[29] While fewer sheep may be part of the way forward, it does not follow that this is the case everywhere in the UK, or indeed in the rest of Europe. González has shown that in Spain "under grazing by livestock or abandonment of pastoral activity is a major threat for the conservation of many types of Spanish habitats".[30] Another noteworthy finding from the Spanish research, given our central concern of timber production is that: "The use of extensive livestock grazing has been proved as a cost-effective conservation tool for preventing forest fires".[31]

Nevertheless, a switch to a more plant-based diet would release large amounts of farmland for other uses including, where appropriate, growing wood fibre, more commercial forestry and more rewilding including native forest expansion. The focus of rewilding would be on halting and reversing biodiversity loss. We shall discuss how we could produce the same amount of food, on *less* land than we use now, when we address the benefits of agroforestry, in Chapter 10.

In the UK the decision by Drax power station to begin to burn a small amount of home-grown biomass (see Chapter 8) is a good example of how land use will need to change and how farmland can be reallocated, with fields that currently grow cereals, rather than graze sheep, around Drax flipped over to a mix of Miscanthus, willow or poplar. The countryside will look different as a result and some people will be outraged, just as in the 1970s when oil seed rape was first introduced into the UK letters of complaint appeared in *The Times*. Now, 50 years later, many people positively like the swathes of yellow which once caused such a kerfuffle.

To conclude, through the glass darkly, we begin to see the outlines of a way forward. To summarise a complex chapter: all land – urban, suburban and rural – needs to be optimized in such a way that we produce the maximum amount of food, forest wood and fibre including biomass for energy. We also need to sequester as much carbon and support as much biodiversity as possible while providing adequate space for healthy recreation. The balance between these priorities will vary from place to place, region to region and country to country.[32] Success will depend on abandoning silo thinking and any hopes of a one-size-fits-all model. We also need to question conventional dichotomies such as wood versus food (see the next chapter). Within this land-use mix, producing the much-needed additional woody fibre will be achieved by bringing into use abandoned and underused land; boosting the productivity of farmland by using wood chips; and by releasing land from meat production for a range of uses, including growing woody fibre. If we undertake these steps, there will be enough land and it will be productive enough to grow the wood fibre we need to help us reach our net zero targets by 2050. Yes, there will likely be a few more angry letters to *The Times*, but, as Stephen Hawking wisely said, "intelligence is the ability to adapt to change".

10

Harnessing the climate benefits of agroforestry

Fact: *Depending on the form of agroforestry new areas of trees on farms can sequester an average of between eight and 16 tonnes of carbon dioxide equivalent per hectare per year during the first 30 or 40 years after establishment.*

Let us recap. The climate crisis we face necessitates two vital responses that are both equally critical and need to be enacted in tandem: an end to nearly all greenhouse gas emissions including CO_2 into the atmosphere *and* the removal from the atmosphere of existing greenhouse gases including CO_2. A major contribution to the first of these tasks can be achieved by substituting the numerous carbon-intensive materials that we currently use in our built environment, including concrete, steel, plaster board, plastic, brick and block, and replacing them with timber-based products that store carbon (see Chapters 1–4). The second task can be tackled in part by planting trees in as many settings and configurations as possible. This will involve improving our sustainable forest management techniques, extending our existing forest cover as and where appropriate, moving closer to harvesting the net annual increment across all European countries (see Chapter 7) and redeploying abandoned farmland for afforestation (see Chapter 9). Together, these steps will increase the amount of CO_2 that is sequestered.

Having increased our tree cover we also need to increase our use of timber as this demand is key to driving the "forest pump", which we discussed in Chapter 2. Ideally as much timber should go into long-life harvested construction materials as possible so that we can store as much carbon as possible in the built environment and reduce the use of carbon-intensive materials. The felled trees are then replaced with new growth enabling the "forest pump" to continue its vital work of sequestering more carbon and then moving it via the timber to be stored in the built environment.

To facilitate and drive the maximum amount of CO_2 emissions removal from the atmosphere via the sequestration attribute of trees we need to increase the percentage of harvested forest wood that is used to manufacture long-life harvested wood products for the construction sector (see Chapter 7); aspire to access sustainable accredited timber from Russia, Belarus and Ukraine when Russia withdraws from occupied Ukraine and peace is restored (see Chapter 7); and switch forest wood from being used for energy to the growing market for construction and renovation products (replacing it with farm grown energy wood) (see Chapter 8). All these new trees – big and small, conifer and broadleaf, young or old, rural or urban – will sequester CO_2 and store carbon. The more trees we have the bigger the carbon sink we will create. The more timber that makes it into long-life products for use in construction and renovation the bigger the carbon sink we can create in the built environment.

The final hurdle we now need to clear is the concern that increasing wood production on farms will reduce food production. This is an understandable concern: surely if more of the farm is redeployed to growing wood then presumably less would be available for growing food? The answer depends on how you plant these trees (see Figure 10.1). If you plant them in a single block then food production will almost certainly drop. But if you adopt agroforestry practices it should be possible to plant the trees while maintaining the farm's level of food production and even increase it.

First a seemingly unconnected segue to consider. It has been recognized for some time that we are not only facing a climate crisis; we are also facing a catastrophic loss of biodiversity. In fact the UK is one of the most biodiversity depleted countries in the world, having lost half its biodiversity since the industrial revolution.[1] Farming practices and urban spread are the two most significant causes. And for that reason it is not a problem unique to the UK and it's one faced to varying degrees throughout the world.[2] At the very least the journey to decarbonize the built environment – as part of the answer to climate breakdown – must not occur at the price of further biodiversity loss. Better still if the journey can help restore, improve and expand our biodiversity, especially on our farms, which cover 69 per cent of the UK's land mass and 38 per cent of the EU's.[3] This loss of biodiversity has involved a loss of tree cover on farms which in turn has resulted in a fall in the amount of carbon stored in tree biomass on farms. Restoring and increasing biodiversity on farms inevitably means planting more trees on farms which consequently will help increase the amount of carbon stored on farms.

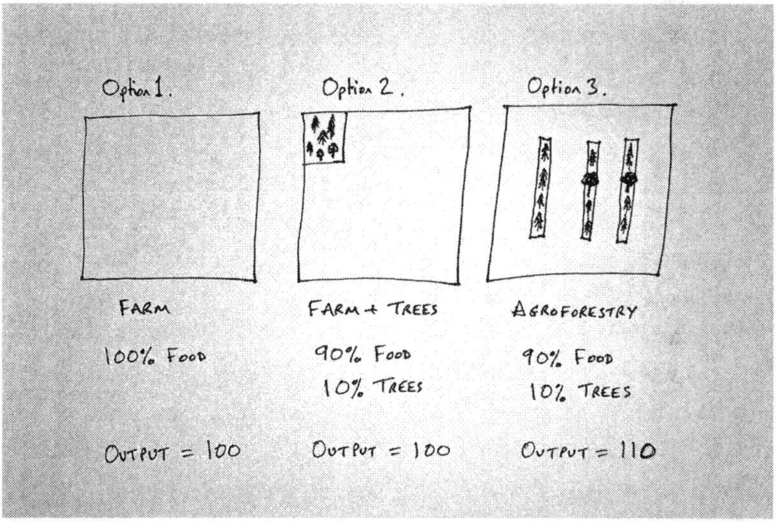

Figure 10.1 Agroforestry can deliver more than the sum of its parts
Source: sketch by author.

There is no such thing as a generic UK or European farm. Rather they come in all shapes and sizes and range from those with very little biodiversity to those which are positively teeming with life – and everything in between. Birdlife is one of the key indicators of wider farm biodiversity, because birds sit near the top of the food chain. The Royal Society for the Protection of Birds (RSPB) have tracked farm birdlife in the UK since 1967 and some of their findings have been downright depressing with the average suburban back garden having a greater variety of birdlife than many farms. Once targeted interventions are undertaken biodiversity starts to climb upwards and it is often the arrival of lost birds – and greater numbers of birds – which is the clearest indication that the decline has been halted and reversed. A growing body of evidence suggests that the biggest single positive biodiversity shift we can achieve on our farms is via the introduction of more trees in the form of agroforestry.[4] The EU Commission defines agroforestry as "land use management systems in which woody vegetation (trees or shrubs) are deliberately grown in combination with crop and/or animal production systems on the same land. Agroforestry has an important role in carbon sequestration, combining significant mitigation effects with co-benefits for ecosystems and biodiversity".[5] Simply put agroforestry is the deliberate introduction of trees into the farm landscape (see Figure 10.2). It can take many different forms and works with both crops and livestock. The chosen tree species could be for timber production and be destined for the built environment, or they could be chosen to produce wood fibre and destined for use in the manufacture of insulation, or for energy wood to replace forest wood, or for food – apples or hazel nuts. Agroforestry markedly increases the range of habitats for insects and wildlife on the farms where it is deployed. *And it need not reduce overall food production.* An extensive EU survey, AGFORWARD, in the late 2010s, detailed the many examples of agroforestry to be found across Europe including pigs and poplars in Italy, walnuts and cereals in Greece, apples and poultry in Holland, willow and cows in the UK, and cherries and maize in Spain. The list goes on and includes crops and trees for timber production.[6]

A European land use survey in 2012 indicated that agroforestry covered the equivalent of about 9 per cent of EU farmland.[7] It used to be much higher but has been in decline over the past 100 years. After the First World War farms began removing their trees while other pieces of land saw the end of farming and became entirely commercial forests. Farmers lost their tree knowledge, previously passed from generation to generation, and a new breed of forester no longer needed to know anything about cereal crops or livestock. I distinctly remember raising the issue of trees on farms at a social event at the National Farmers' Union (NFU) office in Brussels in my first year as an MEP in 2014. The NFU staff member literally took a step backwards and said, "Oh, we don't do trees". I am pleased to report that, (only) ten years later, that has changed. Trees on farms are coming back and for good reason.

Figure 10.2 Agroforesty, Wakelyns, Suffolk, UK - crops and trees
Source: EURAF.

My own experience of agroforestry came courtesy of my French parliamentary colleague Eric Andrieu MEP who encouraged me to visit his constituency in the south of France to observe farms where trees were being (re)introduced. On my arrival the first thing that struck me was that the research which accompanied the tree planting was being paid for by the local water utilities company. The water company was interested in reducing water run-off from farmland into the nearby river, especially during heavy rainfall, as the run-off contained fertiliser, pesticide, herbicide and topsoil – all items which they then had to filter out of the river water before it could be passed on (sold) to customers as tap water. Such run-off is also detrimental for the farmer. It is obviously inefficient to spend money on fertilisers, pesticides and herbicides only for them to be washed away along with that most precious of commodities, the topsoil itself. The water company and the farmers had therefore joined forces to see if the strategic planting of trees on their farmland had the potential to reduce run-off.

Tree buffer strips are an easy option. You simply plant a wide strip of trees along the stream or river edge. Below the trees grass will grow. The combination of the two then acts, in effect, like a sieve preventing larger particles from passing into the river. Such tree buffer strips also slow the passage of water from the land into the river which has become an increasing concern everywhere. Reducing both the speed at which rainwater leaves farmland and the amount can significantly help reduce flooding downstream. For example, since the Boxing Day floods of 2015, 50,000 trees (all sequestering CO_2 and storing carbon) have been planted in the Upper Calder Valley in West Yorkshire, England with the primary purpose of mitigating flood risk downstream in places like Hebden Bridge.[8] The principle at work here is percolation. The tree roots provide passageways that guide the rainwater, assisted by gravity, underground where it is stored rather than entering the river. In similar fashion, the strategic planting of whole fields of willow trees upstream of the River Eden in Cumbria has helped to prevent flooding in Carlisle, with the willow being harvested by farmers every three years to provide biomass for the Iggesund paper mill at Workington on the Cumbrian coast – replacing fossil fuels.

Back to the south of France. In addition to buffer strips immediately adjacent to rivers farmers were also experimenting with strips of trees planted *across* their farmland. The strips were only one tree wide but stretched almost the full length of the fields with space left at the end of each strip for farm machinery to manoeuvre. In other words, the tree strips did not connect to the field boundaries (see Figure 10.1, Option 3). Under the trees, following them across the fields, the farmers grew strips of grass containing wildflowers. The tree strips were planted two combine harvester widths apart. The strips of trees, with the grass strips below, were planted at right angles to the slope of the land. This meant that the tree and grass strips acted as a *series* of buffers to slow rainwater run-off, strip after strip, down the slope of the field. These strips were being planted on both cereal-growing farms (arable) and on livestock farms (pasture). In both cases the tree roots played an important role in helping guide some of the water deeper into the soil and reducing run-off into the river. This water, guided down underground by the roots, then acted as a reservoir which would be available for the cereals, grass and trees themselves when the weather turned drier and hotter.

I asked one of the farmers involved what his neighbour thought of it. "He thinks I'm mad", came the laughing reply. Then his expression changed and in a serious tone he added, "but when I start making more money than he does, he'll copy me – you watch!" These strips of trees – often but not always a central element of agroforestry – also bring other benefits. In the winter the trees provide shelter from cold winds and driving rain, enabling livestock to stay outdoors for longer periods of time, saving on costs including feed. In the summer they provide welcome shade for livestock from the heat of the midday sun, and not only in southern Europe (see Figure 10.3). In addition, the strips of grass below the trees provide a home for insects in turn attracting more birds creating an all-round biodiversity boost. Oddly often overlooked in the calculations the trees themselves can produce or constitute a crop. Apple, pear, plum, cherry, hazelnut, walnut and olive trees, for example, can all be planted, where conditions permit, providing the farmer with an additional income stream. And of course, the focus of this book, they can also produce timber and wood fibre for construction, renovation, insulation, as well as biomass for energy, while not forgetting that as they grow they are all sequestering CO_2 and storing carbon – nature's "carbon storage and capture" in action.

Figure 10.3 Agroforestry: cows and apple trees in Germany
Source: EURAF.

As European summers become hotter due to climate change, we are beginning to see cereal crops in southern Europe halting their growth when temperatures soar and water levels fall. This results in reduced yields. The southern France experience suggests that strips of trees planted across arable land can provide enough shade to allow crops to continue to grow, protecting yield levels (Figure 10.4). This might persuade more sceptical farmers concerned about losing income by taking farm land out of crop production. If hotter summers mean lower yields, then tree planting may offset that loss while also bringing in a wider set of benefits – less run off, more biodiversity, shade for livestock. Hot weather exacerbates run-off because it dries out the soil, hardens it and makes the surface less porous. Then, when the heavy rain arrives – and if there is a slope to the land (and there often is) – the beneficial rainwater runs away and is lost. Strips of trees are very effective at reducing rainwater run-off.

Figure 10.4 Agroforestry: experiment with poplar and barley in Bedfordshire, UK
Source: EURAF.

Anyone thinking that agroforestry is only suitable as a niche activity and not something which mainstream farming could embrace should take a virtual trip to south-west Victoria in Australia to visit Jigsaw Farms[9] run by Mark Wootton and Eve Kantor. In the late 1990s they began planting trees on their beef and sheep farm which covers 3,378 hectares (4,800 football pitches). They now have 25 per cent of their land covered by trees. The majority of this is occupied by a hardwood tree, commonly known as spotted gum, which will create a new income stream as it reaches maturity and is ready to fell. Logic would suggest that with 25 per cent less land for the livestock meat and wool production will suffer a corresponding decrease. Not so. In fact, meat and wool production has remained roughly constant. Overall Jigsaw Farms have seen a net return of 10 per cent plus for 22 of the last 25 years. In addition, biodiversity has soared with 48 bird species in 1996 rising to a remarkable 174 in 2022. As Wootton points out: "It was all co-benefits. The shelter it [the trees] provided increased your lambing percentages, because if lambs get wind-chill they die. With increased shelter, you get warmer soils, which means you grow more grass".

The Jigsaw Farm case study was deemed so positive that the Australian Forest Products Association put Wootton first on a plane and then on a platform at the COP27 UN climate change talks in Egypt in December 2022. They wanted the wider world to see that farmers can significantly increase timber production while continuing to produce the same amount of food *with* an overall income increase.

The UK has relatively little agroforestry, only 3 per cent,[10] which is well below the EU's 9 per cent but it is increasing. Steve Briggs has the largest area of recently established UK agroforestry holding, in Cambridgeshire, England, where he farms 90 per cent organic oats with 3-metre wide strips of apple trees across his fields. These strips are planted with wildflowers below the apples. The flowers attract honeybees which in turn pollinate the apple blossom each spring. An annual RSPB bird count on the farm has documented a steady rise in numbers and species. He used to have one crop, oats. Now he has three: oats, apples and honey, plus the apple trees are sequestering CO_2 and storing carbon. By 2018, nine years after adopting this new approach, Briggs was reporting that the agroforestry was approximately £80–100 per hectare more productive (fruit and cereals) compared to just cereals – at wholesale values.[11] This figure would be higher if the value from fruit into juice were added. After seven and a half years he was making a return on his investment.

A key element of agroforestry is harnessing the power of photosynthesis by having components on the farm that use sunlight during different periods of the year (temporal variance) and in different spaces (spatial variance). As Briggs explains:

> I look at our fields in July–August and the cereal crops are golden and ready for harvest but this means they are not green and photosynthesizing and thus not feeding the soil biota[12] with

> sugars, starches, etc. It's like turning your solar panels off during the sunniest months! The trees on the other hand are in full leaf and harvesting the sunlight and will do so until October–November. They started late in April whereas the cereals were harvesting sunlight all winter and early spring.[13]

In addition, by layering trees, shrubs and crops, the farm gets an added dimension and becomes bigger. No longer only about width and breadth (which is constrained), it is also about height (which is not limited). By managing the right mix of trees and crops by combining size, location, shade, row orientation, etc., you do not need more land to increase production, nor do you lose 10 per cent of your crop by adding 10 per cent trees as you can gain more space by "3D farming".

Meanwhile agroforestry adds significant biodiversity below ground. Deeper and different rooting of trees exploits more soil – depth is as important as height in 3D farming – and improves drainage and biological function. It also helps with soil stabilization, water management and drainage. And as we've seen, it also reduces the loss of expensively applied pesticide and fertiliser.

The soil micro biome[14] based on annual crops is changed by the presence of perennial trees. The biome moves from bacteria dominated soils in cropping to a greater beneficial fungal presence as seen in woodland, including mhychorizal fungi, which is key for nutrient cycling and drought resistance, as well as subterranean pest and disease management. Research is showing that mhychorizal are more prevalent in the soils with agroforestry. All this leads to greater resilience, especially in climate extremes of heat (hot and cold) and rainfall (floods and drought).

After visiting Briggs' farm I was impressed enough to ask for a meeting with Rory Stewart, then a minister in the Department for Environment, Food and Rural Affairs (DEFRA). To his credit Stewart listened with interest and thought a DEFRA visit would be in order. Briggs, accommodating as ever of visitors, agreed. A few months later a pair of DEFRA civil servants arrived in Cambridgeshire. Around the same time Briggs hosted a visit by staff from the Woodland Trust who were joined, following persuasion from Briggs, by the agricultural storyline writer for "The Archers", a long-running BBC Radio 4 soap about rural life in a mythical village somewhere in the middle of England. Shortly afterwards DEFRA started to reference the benefits of agroforestry and an agroforestry farming line appeared in "The Archers"!

The Woodland Trust and the Soil Association are both big advocates of agroforestry and it is worth noting that the *The Agroforestry Handbook* is the Soil Association's most downloaded publication.[15] Groundswell, the UK's largest regenerative agriculture festival, had agroforestry heavily featured at its 2023 gathering. Given that there is an infinite variety of ways in which agroforestry can be applied it follows there are an infinite number of research possibilities. Any one of these could yield further advice on how we can best optimize agroforestry to increase biodiversity levels and our timber production. One such agroforestry research project is being run by the Agriculture Department of Newcastle University. The department has two farms in Northumberland, one in the Tyne Valley and the other, Cockle Park Farm near Morpeth, which is running a trial involving willow and poplar trees. Following a planting pattern advised by Neil Watkins of Eco Crops UK, the trees are planted in 3-metre-wide strips across a cereal crop, each strip has different configurations of trees.

While most willow being planted by farmers is currently being used for energy this could be very different ten years from now. For instance, chipped willow could make its way into the wood panel industry as part of a mix that also included virgin forest chipped timber and recycled, (old) chipped timber. Willow and fast-growing poplar are also sources of wood fibre – the primary raw material of the growing bioeconomy. Willow fibres, like many other wood fibres, could also be used to make clothing, replacing man-made fibres based on fossil fuels, or intensively produced cotton, which demands huge amounts of water in countries that have increasing water constraints. A biofuel for transport can also be made from wood fibre. The wider chemical industry can also extract industrial sugars from wood fibre for use in the pharmaceutical industry. To date, all the evidence from the bioeconomy suggests that everything we currently make from fossil fuels we could make from wood fibre.

Another research project of interest is the previously mentioned AGFORWARD project, steered by Paul Burgess at Cranfield University, which works with a group of farmers in southern England. At an initial workshop the farmers noted their concern about climate change and that they were regularly receiving communications about creating woodlands and selling carbon credits (see Chapter 3). However, they had little idea about how their own farm businesses were contributing to greenhouse gas emissions and how they could be reduced. Although there are a range of farm-based "carbon calculators" available, there was uncertainty about which model to use. Burgess, working with the farmers and students from Cranfield University and using the Farm Carbon Calculator[16] and Agre-Calc model,[17] has found that the net greenhouse gas emissions from an arable farm are typically about two tonnes of carbon dioxide equivalent per hectare per year. If a farm business is seeking to achieve net zero emissions the first priority is to reduce its emissions through the more efficient use of nitrogen fertiliser and its farm machinery. That will reduce emissions to a point, but would be unlikely to be sufficient for a farm business to achieve net zero. It would still be necessary to increase the area of trees growing on the farm. Depending on the form of agroforestry adopted new areas of trees on farms can sequester an average of between 8–16 tonnes of carbon dioxide equivalent per hectare per year during the first 30–40 years after establishment.[18] Hence if annual arable emissions are two tonnes per hectare and newly established trees can sequester 16 tonnes per hectare, then planting an additional 11 per cent of the farm with trees would enable the farm to, on average, achieve net zero emissions over the next 30–40 years. Burgess explains that the beauty of "agroforestry thinking" is recognizing the best way of integrating those trees with food production within the same business.

That planting an additional 11 per cent of a farm with trees opens up a pathway to net zero emissions chimes with the thinking of the European

Agroforestry Federation (EURAF), an umbrella body linking farm, forestry and agroforestry organizations across 23 European countries. Their policy analyst Gerry Lawson argues that a target of 10 per cent tree coverage on all agricultural land by 2040 is the only way that the EU will meet its target of minus 310 million tonnes CO_2 equivalent sequestration by 2030, and greenhouse gas neutrality in the whole of the land sector by 2040.[19]

One of my favourite examples of agroforestry in the UK is that of The Lakes Free Range Egg Company run by farmer David Brass in Cumbria. Here, trees have been planted as a way to tackle ammonia emissions, which is a problem with chicken farms. Trees can absorb ammonia while at the same time providing shelter for the chickens, retain water and reduce run-off. The farm has seen a fall in chicken mortality rates, less disease and better quality eggs. As the trees grow the thinnings provide additional income in the form of biomass and woodchips, while the trees soak up the CO_2 and store carbon. Red squirrels and barn owls have arrived for the first time. David Brass explains the improved financial returns:

> The cost of planting trees is recouped ten times over in the first couple years. A 16,000-bird unit like this one is around about £3,000 a year better off. That's what it would cost to fence and plant this amount of woodland in the first year. So, after the first year every year ongoing you're £3,000 better off. The farm without trees will be a poorer place. If I have to give that advice to other farmers especially with free-range chickens, it's plant more trees.[20]

In Chapter 8 we saw how the Drax power station in North Yorkshire is considering sourcing a small amount of the biomass it burns for energy from farms in the north of England rather than the entirety of its biomass coming from the forests of North America. Initially Drax has focused investigations on Miscanthus (a fast-growing perennial grass) but they have indicated they may support the development of other crops. If Drax were to opt for willow as one of its "bioenergy feedstocks" this would help support the creation of higher levels of biodiversity on the farms where it is planted. Willow flowers annually which is good news for insects, including bees which can feed on the nectar produced each spring.[21] Research by the Department of Zoology at the University of Oxford found that willow scored highest for the "Number of species of phytophagous insects and mites associated with trees in Britain".[22] Oak was in second place. If the insect population grows, then the bird population increases because of the augmented food supply.

Drax is a major player in the world of generating electricity from biomass. Where Drax goes, others are likely to follow. Drax is therefore very well placed to lead the way on two key fronts. The first, a switch from a forest-grown energy wood (biomass) to farm-grown energy wood (biomass, probably willow), would help our forests meet the increased demand for timber from the construction and renovation sectors, especially for wood-fibre insulation, to enable the decarbonization of the built environment. The second, a switch to growing the farm energy wood in an agroforestry setting rather than as a monoculture (a single block of forestry on 10 per cent of the farm), would significantly increase the biodiversity to be found on our farms.

Agroforestry offers farmers not just emissions solutions but the potential for better financial returns. For it to work, farmers need to calibrate – using robust science and commercial evidence – which combination of livestock and trees, or crops and trees, works best for their land and their location. The fact that there are so many permutations is both attractive and intimidating. If they get it right, then it can drive up their income. At the same time, and with little additional effort, it will increase the biodiversity on their farm, benefiting us all. It is also likely to help generate income from future government subsidies, but only if we develop a system of payments that rewards experimentation and evolution. Crucially *the trees, whatever type they are, will absorb carbon and store it.* This stored carbon will also make its way into the built environment if the trees are converted into wooden construction products or wood fibre insulation.

Why then if agroforestry offers so many benefits, especially with regards climate change mitigation, has its take-up by UK farmers remained to date low? This was the question a recent report funded by DEFRA and conducted by a team of progressive UK farming and woodland advocacy organizations set out to answer.[23] Synthesizing ten of the best recent surveys and reviews on agroforestry adoption in the UK the researchers found that nine of the top ten disincentives to UK farmers adopting agroforestry involve either a lack of technical knowledge or certainty about the financial returns.

With the uncertainty of post-Brexit agricultural policy in the UK many farmers are hesitant to commit to long-term investments in agroforestry without knowing what exactly will be incentivized. Clearly there needs to be explicit funding made available to encourage agroforestry planting, yet such funding should also be flexible to allow systems to be designed for the context of the farm on which they are being established. DEFRA have suggested an agroforestry standard will be included in the Sustainable Farming Incentives while the Organic Research Centre and others are coordinating a series of trials to understand what this standard may entail.

Agroforestry, like many other agroecological endeavours, is knowledge intensive. Furthermore, the type of knowledge required is not merely theoretical but is grounded in the contexts of the land and farming systems in which trees are to be planted – what may work on one farm will have very different outcomes on another. With only 3.3 per cent of the UK's land containing agroforestry systems there is a limited supply of inspiring success stories available for British farmers to build upon. The research rightly highlights the need for more UK-specific case studies and agroforestry demonstration farms. As the report states: "Farmers just don't feel they know enough about agroforestry to go ahead and plant an agroforestry system. They also feel it is a substantial long-term financial investment and don't feel the current incentives and payment systems are sufficient".[24]

The report also received input from the Soil Association: "Farmers are hungry for more technical information on how to implement and manage agroforestry systems, as well as wanting a better understanding on the financials of including more trees on their farms".[25] And the Woodland

Trust's farming lead, Helen Chesshire, was positive but also concerned: "I believe there is a strong argument for including an explicit agroforestry component in the main farmer payment system to have emerged post-Brexit: the Sustainable Farming Incentive. As it stands currently, I see a distinct possibility of agroforestry falling through the cracks of post-Brexit agriculture policy".[26] Briggs is optimistic, "The benefits of agroforestry for both the productivity and diversity of my farm have been immense. Farmers are aware of the term 'agroforestry' now and I think there is a very strong argument for a more distinct inclusion of agroforestry into UK agricultural policy post-Brexit".[27] There is a distinct sense of change in the air. Ian Brown, a former tenant farmer in Northumberland and early adopter, describes it this way, "Agroforestry is gathering momentum. The tipping point will happen when farmers realize the old way of financing agricultural activity on land have changed. Policy and funding will provide the bridge – once outputs are measured and rewarded".[28]

I would add a final benefit of agroforestry – beauty. Having visited numerous examples of agroforestry across Europe, including in the UK, I believe that agroforestry adds a welcome variety to the rural landscape. It can look different from the traditional rural idyll – there is no getting away from that. Also its rapid and extensive introduction, which is what is needed to tackle both climate change and increase biodiversity, will no doubt be met with grumbles, complaints and objections from some. However, it can be beautiful and aesthetic sensibilities regarding the countryside have changed many times over the centuries.

It is the significant impact that a move to 10 per cent agroforestry on EU farms could have on the amount of CO_2 sequestered and carbon stored that is the key consideration. Given the EU's current forest cover and its farm cover are about the same, at 38 per cent, then a 10 per cent increase in tree planting on all EU farms would roughly equate to having increased the capacity of the EU's forests to sequester CO_2 and store carbon by 8 per cent.[29] Such a change would increase the current annual CO_2 sequestration figure of 155Mt by an extra 15Mt to 136Mt. This would result in the EU's tree cover moving from offsetting 10 per cent of the EU's annual carbon emissions towards offsetting 11 per cent annually. The maths for the UK would be even better as we have a higher percentage of farmland at 70 per cent and a much lower percentage of forest cover at 13 per cent. Consequently a 10 per cent agroforestry figure would – from a CO_2 sequestration and carbon storage perspective – have the equivalent impact of having increased UK forest cover to 16 per cent. Done in conjunction with a steady rise in the amount of carbon stored in construction and renovation products in the built environment, especially insulation and the reduction in the use of carbon-intensive materials such as concrete, and you can see the potential wood has to help save us from climate breakdown.

To conclude, effectively tackling climate breakdown often throws up hard choices and difficult trade-offs, consequently achieving any progress always seems to have a downside. Agroforestry appears to be an exception. To decarbonize the built environment successfully we need to grow more sustainable wood. We will struggle to gain public support (and hence the crucial political support) if increased tree planting on farms were to reduce food production and weaken our food security. But it does not need to. Likewise, we would struggle to gain support for increased tree planting on farms if doing so reduced biodiversity levels. But, it does the opposite – it increases biodiversity. The evidence is steadily mounting that planting just 10 per cent of a farm with trees as agroforestry (hence spread across the farm), not as a single large new wood (concentrated in one place), achieves the multiple wins of an increase in CO_2 sequestration and carbon storage, no reduction in food production, a major boost for biodiversity, the development of a new income stream and an increase in returns for the farmer.

And returning to the message of the earlier chapters, when the trees are harvested and leave the farm, they can go on storing carbon as construction products in the built environment, specifically wood-fibre insulation, or they can be used as energy wood acting as a replacement for the forest wood that is currently burnt. Forest wood that could be more optimally deployed as a construction material or as a raw material for any one of the growing number of products that are being developed by the wider growing bioeconomy. Agroforestry is therefore at the heart of a set of interconnected land-use practices that are necessary to provide us with the sustainable timber we need to decarbonize our built environment.

11

Factory-built houses

Fact: *in 2022 in the UK 1 in 60 new homes was made in a factory and the ratio is set to rise.*

Cars are expensive to buy and it would be pretty shocking to discover they were assembled outside in all weathers. Homes are a great deal more expensive than cars and yet we do assemble them outside in all weathers – at least until the roofs go on and the windows are installed. The consensus seems to be that houses, offices, schools and all our other buildings are built the way they are because it is simply what we have always done. However, *how* we build is set to change and wood is at the heart of this possible change. This chapter will look at the potential to build homes like we do cars, *in factories*, and deliver them direct to site essentially ready, or almost ready, to inhabit. As wood is ideally suited for these factory builds – and is therefore the main construction material used – and our aim is to use more timber in the built environment for all the excellent reasons we've outlined, changing the way we build to one that can facilitate that is something we need to actively support.

Fans of the Channel 4 "Grand Designs" television programme may remember the 2004 episode which followed retired couple Greta and David Iredale all the way from Walton-on-Thames in England to Hartenfels in western Germany to see the "HUF HAUS" they had ordered from a catalogue being manufactured in a factory. The HUF HAUS company, active since 1921, are the "gold standard" of flat-packed, delivered-to-site, ready-to-assemble homes. Seen by many as the epitome of luxury and exclusivity, they are iconic structures of timber and glass that do not come cheap (see Figure 11.1). This exclusivity is maintained by a production limit of 200 houses per year, 40 per cent of which currently head to the UK.[1]

Figure 11.1 A timber-framed HUF HAUS manufactured in a factory in Germany
Source: HUF HAUS.

At the other end of the price spectrum are Bristol-based Agile Homes who specialize in small one-, two- or three- bedroomed timber-framed houses, still with style.[2] They are aimed at filling a specific need, usually in an urban setting, with a new home for homeless people or those on low incomes. A consortium of experts identify an unused or derelict piece of land which is suitable for an Agile Home and they then set about raising the funds to have one made and delivered to site. Alternatively, these homes can be bought outright, part-owned, leased or rented individually, or in multiples, for larger families and larger sites. Just as the finished homes are delivered to site by lorry, they can be removed and driven away if needs be. This means that they are technically mobile and hence they do not require the same degree of planning permission as a traditional build.

They can even be designated as "temporary" under the UK's Caravan Act of 1968. You can also buy an Agile home as an annex for an existing home to accommodate, say, an elderly relative who wishes to maintain a degree of independence. As an enlightened company Agile have deliberately engaged local prisoners as staff helping them to improve their skills and provide certified training and education in modern methods of construction in preparation for their release, and potentially to build the homes that they themselves might end up living in once they are back in the community.

At the heart of Agile Homes' design is ensuring that their construction is climate friendly. They achieve this by using timber for the structure and straw for the insulation. This results in lower carbon emissions and lower energy costs compared with traditional building methods. Their homes are warm in the winter and cool in summer (see Figure 11.2).

Figure 11.2 A timber-framed Agile home made in a factory in Bristol, UK
Source: Agile.

The smallest HUF HAUS sells from around £400,000 plus the cost of the land and the foundations. In part this is a reflection of exceptional attention to the tolerances and assembly details that enable a better performance of the building as a whole. A basic three-bedroom Agile Home sells for around £160,000. There is obviously not much in common when it comes to price but they are both made in a factory setting and both have opted for timber frames, the reasons for which we will discuss shortly.

A house made in a factory is often referred to as "prefabricated". For some people prefabricated homes may trigger thoughts of postwar temporary housing which went up across the UK in the late 1940s in order to quickly replace bomb-damaged homes which were uninhabitable. The Emergency Factory Made Housing Programme was announced by Winston Churchill during one of his famous radio broadcasts in March 1944 and its establishment led to the construction of 156,000 prefabricated homes. Factories which had been making armaments were reassigned to making houses which could be delivered to site by lorry and assembled quickly – much quicker than building brick by brick. The prefabrication idea had come from the USA and, for those bombed out of their homes, these "prefabs" were a godsend.

Churchill's government issued a core design brief including maximum section dimensions to allow transport by road. The materials used, however, varied in quality and included aluminium left over from aircraft production, steel or timber frames with asbestos cement sheets, and timber frames with pre-cast concrete sheets. These single-storey detached homes with low pitched roofs were designed to last up to ten years (see Figure 11.3). However, some 70 years later a few remain across the UK and are still inhabited, often loved, by their occupants. They were all built with an inside flushing toilet and bath, running hot water from a back boiler behind the coal fire, a rudimentary central heating system and large glazed windows, providing increased light. For the majority of "prefab" occupants these were improvements on the flat or terraced house with a toilet in the backyard which they had left behind. Another improvement, in a postwar world with rationing, was that they all had gardens where vegetables could be grown. Not surprisingly at the time they were seen by many as the height of luxury.

While assembly was indeed quick it turned out that the first prefabs were more expensive per unit than the brick-built homes they sought to replace. This, along with quality issues given that they were only designed to last up to ten years, led to the end of the scheme by the early 1950s. However, a second and much larger prefab building wave quickly followed led by councils working to tackle the postwar housing crisis. These homes were designed and built to be permanent and were mainly two-storey. They were made from precast reinforced concrete and consisted of concrete panels reinforced with steel, bolted together, often in conjunction with a steel frame. Like Churchill's prefab predecessors they were factory-made and delivered to site by road.

Figure 11.3 1940s prefabricated housing in the UK
Source: Imperial War Museum, London.

While a politician I canvassed in areas which included some of the original remaining prefabs. They still existed across parts of northeast

England in places like the Springwell Estate in Sunderland and I was struck by the strong sense of community that existed amongst the prefab occupants. The original prefabs along with their second-wave cousins, however, are not noted for their aesthetics or energy efficiency, built as they were during a period when energy was deemed "too cheap to meter". These two deficiencies along with issues to do with damp and structural problems, meant that by the 1980s prefabs had a poor image in the minds of the wider public. This is why the latest factory-produced homes are deliberately not referred to by their manufacturers as "prefabricated". Instead, the terms "modular homes" or "modular off-site production" are used. You will also hear the term "modern methods of construction" (MMC) being used particularly at a UK government level. This essentially refers to a process which focuses on off-site construction techniques, such as mass production and factory assembly, as alternatives to traditional building. The process has been described as a way "to produce more, better quality homes in less time"[3] – a description that goes a long way to explaining why there is once again interest in making homes in factories. It should also be noted that MMC lend themselves to subsequent disassembly and reuse of the structural component parts of the build (see Chapter 7).

I have attended many presentations about modular house construction and have noticed that often speakers use a 2017 graph produced by the McKinsey Global Institute, which shows the gross value added, per hour worked, in various industries since 1947 in the USA.[4] The message the graph spells out is clear. Since 1947 agriculture has seen the biggest, fastest rise in productivity for hours worked. Fewer farm workers now produce more food because of innovations involving fertilisers and pesticides (not necessarily a good thing), higher-yielding seeds, plus larger and more powerful farm machinery. Manufacturing is in second place due to the automation of many factories. For example productivity in the car industry has soared as it has moved away from the assembly of single cars to constantly flowing production lines enabling the likes of Nissan to "build" one car every minute at Europe's most productive car plant in Sunderland. Wholesale and retail have followed a similar upward trajectory with the arrival of the supermarket which needs fewer staff than a traditional shop, has barcodes on products for enhanced stock control and provides self-check-out where the customer replaces paid staff.

Conversely, gross value added for the hours worked by employees in the construction industry has flatlined for a staggering *70 years*. In other words, *you need as many people to build a house today as you did in the 1940s* which in turn has helped keep house prices high. There is a marginal upwards rise in productivity (then fall in the 1970s) and the joke is that this is accounted for by the invention of the nail gun in the 1950s, which finally caught on some 20 years later. As other sectors have seen significant increases in productivity, the construction industry has lagged behind: "Despite the development of the construction industry's skills base in terms of increased qualification attainments and participation levels in training, this has not translated into concurrent improvements in productivity performance over the period (1995–2006)".[5]

In 1988 small- and medium-sized firms built 39 per cent of new homes in England. By 2020 this figure had dropped to 10 per cent. The 2008 global financial crisis was a major contributing factor, but the decline began long before. Conversely the top ten housebuilders grew in size and now dominate the market. One impact of this reduction in competition has been a lack of innovation in house building. In response the UK government via Homes England are looking to see how small- and medium-sized building firms can be encouraged to return to the market thereby increasing diversity of supply and more competition.[6]

Housebuilders can make big profits and therefore – many believe – have no incentive to lower their prices. And while enough people manage to scrape together just enough money to climb onto the housing ladder in the UK, even in London and the southeast of England where house prices are 14 times the typical household income, house builders will continue with their existing business models. What the house buyer needs is the housing equivalent of the Model T car, assembled on a production line and offering a decent high-quality home at a price below that of the traditional house.

There is a remorseless logic at work here. A traditional building site is open to the elements. When it rains everything gets wet and flooding is likely especially when the foundations are being dug. Bad weather delays production every year on building sites. In winter months the working day is constrained by only 7–8 hours of daylight, although major building sites in cities can be floodlit to extend construction hours. However, in the suburbs – where most homes are built – sites become dark and work stops when the sun goes down, not least for good safety reasons. Bringing production indoors into a well-lit environment solves the problems of bad weather and lack of daylight. Shift work now becomes a possibility – as in a car plant – opening up 24/7 production which in theory can open the door to increased productivity.

According to the UK Manufacturer's Organization: "Modular housing . . . has matured rapidly in the last 5 years [2017–2022] – more than doubling its output despite the Covid-19 pandemic. In 2022, one in every 60 new homes in the UK will be built entirely in a factory – that's over 3,300 new homes, providing somewhere to live for an estimated 8,000 people".[7] Modular homes can either be delivered "flat packed", HUF HAUS style, by lorry to site where the separate walls are assembled and a roof added – two dimensional (2D) – or four walls can be pre-assembled in the factory, with a floor and ceiling or roof, creating a modular unit. This is then placed on a low loader and delivered, static caravan style, to site – three dimensional (3D). The latter building, allows for everything to be pre-fitted in the factory – plumbing, electrics, tiling, kitchen and bathroom units. The majority of construction and assembly takes place in the factory meaning considerably less working time spent outside on site (see Figure 11.4). Window glass is one of the few elements that can be problematic if fitted in a factory as panes can crack when the modular homes are transported.

The primary constraining factor with 3D modular off-site construction, where the walls are already assembled into units, is the dimensions of loads which can be transported by road. Too high and it will not go under a standard bridge; too wide and it will stray dangerously into the next lane of the motorway; too long and it will not go around a 90-degree corner. As any motorway user knows, we do see very large items being moved by road from time to time but this normally involves an escort and tailbacks. Delivering hundreds of huge modular home units by motorway would not be popular with other road users. Moreover, the added expense involved (for instance paying for the escorts) would negate the financial savings made by factory assembly. Therefore, the dimensions 3D-modular house builders and architects need to work with are currently akin to those of the largest static caravan – 12 metres × 4 metres × 3 metres high.

Figure 11.4 Inside a modular 3D homes factory in the USA
Source: US Department of Agriculture.

Another downside of 3D-modular builds is that you are moving large amounts of fresh air from factory to site, i.e. there is nothing much in most of the rooms you are transporting with the exception of fitted kitchens and bathrooms. This pushes up the transport costs in comparison to 2D-construction where the lorry loads are mainly all house and little air. Additionally, 2D is released from the strict room-size constraints of 3D. The single 3D units are also heavier as individual items than the multiple individual 2D-wall units, which can necessitate a bigger more expensive crane to assemble 3D units on site. As Agile's CEO and co-founder Craig White explains it:

> Agile's model is called Distributed Manufacture, which unlike 3D has no requirement for large capital investment and instead works with existing supply chains to deliver 2D flat-pack MMC. The cost differential between the two approaches is substantial with 3D having eye-wateringly expensive transport and cranage costs, plus the need to recover investment into large factories. 3D volumetric has a key restraint on the shape of rooms because it must be transported. This results in longer and thinner rooms. Housing should not be designed to solve a transport constraint that represents a tiny fraction of the life of a home. After all transportation and installation may account for only one week of a minimum 60-year house life. A house is a home, not a car.[8]

One company that had achieved success with modular homes was Urban Splash which began life converting old warehouses into flats in inner-city areas, such as in Manchester. They then diversified into factory-made modular 3D units. The units, two or three storeys high, allowed the buyers to decide whether to have the bedrooms on the ground floor, the kitchen on the middle floor and the living room on the top floor – or any variation thereof. Their pleasantly curved crescent of homes in North Shields, overlooking the River Tyne at Smith's Dock, is a typical example of their approach (see Figure 11.5). As of 2022 they had built 1,000 modular homes. Tucked in behind Urban Splash's Smith's Dock crescent is a shorter row of two-storey modular timber-framed homes, branded "Fab House", the brainchild of TV architect and northeast born and bred George Clarke working with TDO Architects.[9] Their outside appearance is essentially neutral but inside the beauty of the exposed wood is emphasized. For Clarke modular construction is one way to raise the standard of UK homes: "I believe that in a country that unfortunately has too many poorly designed and badly-built homes, this was an obvious route to take".[10]

Figure 11.5 Prefabricated 3D Urban Splash factory-built homes at Smith's Dock, North Shields, UK

Source: Paul Richter.

Material wastage is notoriously high on outdoor construction sites with most of it ending up buried as landfill. Manufacturing houses in factories means reduced waste. A modular homes production line uses standardized items that have been manufactured in a supplying factory or even in the same factory. Higher accuracy levels are achievable and, in this way, everything fits together on the production line with little or no waste. This is not dissimilar to a car assembly line. The highly productive, efficient and profitable Nissan car plant at Sunderland does not really *make* cars, rather it *assembles* cars from hundreds of individual parts made elsewhere, which are then delivered "just in time" to be put together on a continually moving production line. Additionally, factories are safer places to work than building sites with the construction industry regularly recording fatalities each year.

Factory-based house production also has a positive social impact. While visiting one of the small modular homes factories just outside Helsinki, I noticed that the workforce included a small but noticeable number of women, just like the Nissan car plant in Sunderland. This contrasts with traditional building sites where there are few women workers. I asked the deputy manager who was showing me around the Helsinki factory about the female staff. "We work a shift pattern and we can accommodate childcare roles by staggering when we start and finish shifts, something which external building sites can't do because they're constrained by daylight".

This discussion of factory-built houses may be interesting, but how does it fit in with the argument that timber can save the world from climate change? The answer lies in the materials used. The majority of homes in England have a frame made from individually-laid bricks on the outside and blocks on the inside. This technique is not feasible for a moving production line producing factory-built homes that then are transported to site. Therefore, the choice for architects of today's modular homes is whether to use a wooden, metal or concrete frame to assemble the house around. Clearly, from a climate perspective, if modular homes can be built with timber frames then this is far more beneficial. But can they?

The key issue concerning materials and modular homes is *weight*. The heavier the materials used the more expensive it becomes to deliver the homes to site by lorry and the more expensive to assemble them by crane (see Figure 11.6). Lighter loads mean less diesel burnt, meaning lower costs. As timber is the lightest of the three classic building materials (wood, metal, concrete) it presents a clear financial advantage. Consequently, timber has become the dominant structural material used to make modular factory-built homes because its relative lightness of weight brings down delivery and assembly costs.

In addition, lighter loads reduce both CO_2 emissions and air and water pollution.[11] Of the four main particulates which damage air quality one comprises the tiny fragments of rubber which are released from vehicle tyres, heavy goods vehicles being the major culprit (see Chapter 4). Although neither of these "costs" are paid for by transport companies, it is a price we all pay.

Figure 11.6 Crane lifting wooden modular units into position to create an office block in Berlin
Source: iStock; JARAMA.

This move towards the prefabricated factory production of modular homes has not however been all plain sailing – far from it. The costs involved, in particular to set up a major factory operation on a scale similar to a car plant, are significant. Deep pockets are needed along with the ability to hold one's nerve as the costs rack up and the returns seem ever distant. In 2022 Urban Splash's prefabricated operation, known as "House by Urban Splash", went bust with the loss of 160 jobs at its Derbyshire factory and at other sites across the UK. Administrators blamed "the under-performance of its modular facility" due to "design issues resulting in production defects and re-working the modular units, the cost of which could not be passed on". In 2016 Legal & General (L&G), one of the UK's biggest insurance and pension firms, began making modular homes at a huge factory at Sherburn in Elmet – midway between Leeds and York and with excellent connections to the UK's motorway network. It has since struggled and ceased production in May 2023 giving the reason, "not been able to secure the necessary scale of pipeline to make the current model work".[12]

Clearly, the failure of L&G's operation raises important questions. Was it the right model? Would 2D have worked better than 3D? If it was the right model, why didn't it work in a UK market crying out for affordable homes? Did existing building regulations help or hinder? Did the sustained month on month rise in the cost of borrowing destroy demand? Why was the whole operation so resource-intensive requiring huge capital investment? Why did they use cross-laminated timber (CLT), a product best suited for building multi-storey buildings, rather than an LVL or simple timber frame, which would have been cheaper and more resource-efficient? While it may be too late for L&G, if we don't seek answers to these questions we will end up trivializing the decision-making process that led to this major investment and not offer solutions for other players entering the market.

An existing issue, reinforced by L&G's cessation of production, is the lack of a feedback loop in the UK construction industry that enables the assessment of performance in a quantitative manner. This results in many failures in traditional construction (often attributed to failures of "workmanship" but typically resulting from a lack of well-defined targets for performance), as well as many untested regulations that limit the efficiency of construction. If you have well defined quantitative performance targets you will develop the technologies to meet them in the most efficient manner. The construction industry lacks these and this issue is a major constraint in industrializing construction.

What we are learning is that while the concept of modular off-site construction may be simple (and attractive) its application, however is not. As observed by Mathew Aitchison, Professor of Architecture at Monash University, Australia: "It has cast a cyclical 'spell' over many in the industry: each successive generation to rediscover prefab applies prodigious amounts of energy to it, only to encounter similar problems and mistakes".[13] Likewise, Jose Torero Cullen, Head of the Department of Civil Engineering & Geomatic Engineering at University College London, strikes a similar precautionary tone: "Any form of 'factory made' housing needs to address a very complex problem, predicting the system performance once it is assembled on site. Predicting component performance when it comes to aesthetics, structures, thermal, sealing and so forth is very easy, but addressing the same problems during assembly and characterizing performance remains the key challenge".

One construction industry innovation expert who anticipated the demise of the modular operation of L&G was Robert Hairstans of the UK's New Model Institute for Technology and Engineering: "When I visited their factory last, I simply could not believe what they were doing and it genuinely was just a matter of time. It's sad because it undermines the factory-based approach because it's simply all wrong".[14] According to Hairstans, L&G failed partly because of the inherent problems that exist in the construction sector, ultimately stemming from its fragmented structure, culture and complexity. But also because the concept, hugely capital-intensive and over-engineered, was trying to shoehorn volumetric CLT boxes into the low- and medium-rise UK standard house typologies – a complete mismatch. Hairstans points out that successful off-site manufacturing operations, of which there are many, have predominantly innovated incrementally and invested in their delivery approach over decades. House typologies have evolved to be compatible with the techniques and technologies deployed, he argues, and they have responded to market drivers such as enhanced energy performance and modern living requirements, such as open plan living.

Does the L&G experience spell the end of large-scale 3D modular? No. In 2022 two new companies – Nokera and TopHat – both entered the modular homes market.

Nokera, although headquartered in Switzerland, are a German company formed in 2021 with the expressed aim of becoming "the world's largest manufacturer of wooden apartments and houses". They intend to build no fewer than 20,000 2D- and 3D-wooden housing units a year across four production sites in Germany, their main factory being at Möckern near Magdeburg in Saxony-Anhalt. Nokera may well be on safer ground than L&G as Germany has been making more modular homes for longer than the UK, as have Switzerland, the Nordic countries and Japan. This point is significant as these are all environments that promote "industrial thinking" with manufacturers, with regulators and with the public, and it's no coincidence that both Germany and Japan have a well-established reputation for high-end manufacturing as seen in their respective car industries.

Like most of their predecessors Nokera have opted for timber frames, as they explain:

> Germany is short of 1.5 to 2 million housing units in the middle and lower price brackets. The goal is to construct an additional 400,000 sustainable housing units per year. And the only way to achieve that goal is through innovative solutions.[15] *Building with wood* makes it possible to produce environmentally friendly housing quickly and cheaply. At NOKERA, we are transforming the construction process for affordable housing with a resource-efficient approach and a specially developed digital process chain. Our completed NOKERA buildings *will store large quantities of carbon for a long period of time.*[16] (emphasis added)

While HUF HAUS will go on building their Porsche standard modular homes Nokera have arrived to build the Volkswagen equivalent.

At the same time as Nokera were limbering up to enter the German market TopHat were doing the same in the UK. TopHat is funded and 70 per cent owned by Goldman Sachs, the US based investment banker. Their website claims that "TopHat was founded to deliver beautiful homes, sustainably. We use advanced manufacturing and the latest technology *to deliver ultra-low carbon homes*, whilst keeping costs as low as possible" (emphasis added).[17] That TopHat are thinking big is evidenced by their initial budget – £130 million – and their plan to build a factory the size of 11 football pitches in Corby which will provide 1,000 highly-skilled jobs. If, as claimed, the factory will be capable of building one house per hour, it will be Europe's largest modular homes facility.

Watching developments closely are the UK's big traditional-method housebuilders. They may have breathed sighs of relief when early mover Urban Splash ceased modular production followed by – more significantly – L&G, but the appearance of TopHat must have been a shock. For some time there has been speculation as to which of the existing big housebuilders would jump first into the deep-end of modular homes. In March 2023, we were given the answer. Barratt made the following headline in *Housing Today*: "Britain's biggest housebuilder Barratt is to open a £45m timber frame factory to build more homes off site. The firm believes the factory will help it reach its *commitment of becoming carbon neutral* by 2030, by which time it wants 30 per cent of its homes to be built using modern methods of construction" (emphasis added).[18]

It would be difficult to understate how much pressure this then put on the other major UK housebuilders to follow suit. They certainly noted Barratt's publicly stated intention of using a timber frame system that will enable them to cut the build time on their housing developments by an average of five weeks, compared to the traditional masonry method (those carbon-intensive bricks and blocks). Housing insiders will have noted that this five-week difference will in turn clear the way for an all-important cheaper house price for the customer.

Barratt were then followed by rival housebuilder Persimmon who in May 2023 teamed up with insurance company Aviva to make a £70 million investment in TopHat – so clearly the demise of L&G's modular operation has not put others off. Persimmon's investment is significant as it occurred *after* the demise of L&G's operation. L&G had been the first FTSE-100 company to invest directly into modular, Persimmon are the second.[19]

These moves by Barratt and Persimmon will now put pressure on other major UK housebuilders to follow. They will have also noted Persimmon's chief executive stating: "This investment provides Persimmon with guaranteed access to very energy-efficient volumetric modular units. This will provide further build efficiencies, manage the growing challenge of labour shortages in key trades and expand our product range for customers".[20]

It appears that the game is afoot: has Henry Ford, the pioneer of mass assembly-line production finally driven into town? For every construction expert who thinks he has, there is another who thinks he hasn't, at least not yet. And do we truly understand how to bring the "factory" approach to construction?

Despite these recent stop-start developments in the UK modern modular house construction in the British Isles lags well behind the Nordic countries. IKEA, the home furnishing brand, has been erecting modular homes in Sweden, Finland, Norway and Denmark since 1997 via a subsidiary company named BoKlok – a joint venture with the construction firm Skanska. To date they have built around 14,000 homes with 90 per cent of the construction occurring indoors in a factory setting using wooden frames and MMC.[21] BoKlok homes are not 2D flat-packed despite their link to IKEA – rather they are the 3D-modular units as used by L&G and Urban Splash. "Clipped" together on site BoKlok manufacture all their homes using wood because, in their words: "[Wood is] the most sustainable building material, with the lowest emissions, compared to others. BoKlok homes are completed off-site in a safe and dry environment using a smart and industrialized process. This means that we can reduce waste and energy usage in the factory".[22]

BoKlok have now entered the UK market. Worthing and Littlehampton, neighbouring towns on England's south coast, are set to be the first locations for "IKEA" homes in the UK. Initially BoKlok's UK homes were constructed at Urban Splash's factory in Derby, but since the demise of House by Urban Splash the factory has been bought by TopHat who will now supply BoKlok. TopHat have also opted for timber frames for *all* of their modular units, a cornerstone of their ultra-low carbon approach.

Tipping points are a peculiar phenomenon. Few see one coming but once they have happened, everyone can see that they were inevitable.

Thirty years ago our high streets did not normally contain coffee shops; now they are awash with them. Thirty years from now will may look back and ask ourselves, "Did we really build houses outdoors in all weathers?"

Two criteria will decide the fate of modular, off-site construction, although it should really be three. First will be quality: can modular homes gain a reputation for being built to a higher standard than on-site construction? Second is price: can modular homes be sold at a lower price than equivalent homes built on-site? If these two targets can be achieved McKinsey will need to redraw their "value-added" graph in 20 years' time.

The third criterion which really should be included – but will not be given current approach to house building – is to do with carbon emissions: can modular homes help drive down carbon emissions from the built environment? The answer to this question is a resounding "Yes". This is because they will, when built with timber frames and wood-fibre insulation, increase the percentage of homes with significantly lower carbon footprints from their construction (emissions reductions) *and* they will be making a major contribution to our capacity to store carbon (emissions removals).

Conclusion: Timber Rising!

> *"Anyone who has a new idea is a nutcase until the idea catches on."*
>
> *Mark Twain*

> *"We all know that today building with timber could save up to 40% of carbon emissions in comparison to concrete. By keeping the carbon inside the wood, one day timber could turn our homes and even entire cities into carbon sinks."*
>
> *Ursula von der Leyen, President of the European Commission, Nordic Bauhaus Conference, 24 November 2022*

As the apartheid regime in South Africa entered its death throes the octogenarian president of the British Anti-Apartheid Movement (AAM), Archbishop Trevor Huddleston, became increasingly worried that he would die before Mandela was freed, apartheid was vanquished and a new democratic South Africa was born via the ballot box. He therefore took to haranguing, and only in a semi-friendly way, the staff of the AAM when he visited us at the movement's headquarters in an old piano workshop in Mandela Street, Camden, London. To help reinforce his key message that we all needed to work harder he always told us the same, short, humorous tale. On a "recent" visit to his spiritual home – the College of the Resurrection at Mirfield near Leeds – he recounted that over breakfast he and his fellow octogenarians had fallen into a conversation about the pros and cons of buying a new suit at such an advanced age, whereupon the oldest member of the community, who was in his nineties, piped up "New suit? New suit? I worry about buying green bananas".

When it comes to climate breakdown we are without question in "green banana" territory. There really is very little time left. In the words of António Guterres, the Secretary General of the United Nations, "Climate change is here. It is terrifying. And it is just the beginning. It is still possible to limit global temperature rise to 1.5°C and avoid the very worst of climate change. But only with dramatic, immediate climate action".[1]

As we seek adequately to address the threat we face, there is no longer any room for manoeuvre or any time for prevarication. In the past we possibly had the option to do a little less of this by doing a little more of that. But those days are gone, even if many people appear to be behaving otherwise. Now our only credible course of action is to undertake all the identified climate actions, and immediately. All possible activities both to *avoid* greenhouse gas emissions and *remove* greenhouse gases from the atmosphere need to be undertaken now. It is in this context that *Timber!* was written. How do we ensure that wood makes its maximum contribution to addressing climate breakdown?

We began in Chapter 1 by identifying the problem that the use of timber can help to address: our built environment is currently responsible for around 40 per cent of global CO_2 emissions. The majority of this (29 per cent) is due to the burning of fossil fuels to heat and cool our buildings. We can significantly reduce these emissions by insulating *all* our buildings to a very high standard (Chapter 3) and by switching our energy consumption away from fossil fuels to renewables. A smaller but still significant amount of emissions come from how we build (11 per cent) and are the result of the materials we use to construct our buildings, with concrete being the standout problem material, responsible for 8 per cent of global carbon emissions. Consequently, in Chapter 2, we began the process of exploring timber's ability to *substitute* for the materials that have the highest levels of embodied carbon. This is the area where the academic evidence continues to mount: "Several recent studies have shown that substituting mass timber for steel and concrete in mid-rise buildings can reduce the emissions associated with manufacturing, transporting, and installing building materials by 13–26.5%".[2]

While it is self-evident that insulating our buildings to a high standard drives down energy use, which in turn often leads to a reduction in the burning of fossil fuels, what was less clear, at least to begin with, was which materials should be used to make the insulation. By the end of Chapter 3 we concluded that potentially all insulation materials – almost irrespective of the material they are made from – could have equally high levels of embodied energy. However, only the *nature-based* insulation materials could *store* carbon. This ability to *remove* CO_2 from the atmosphere means we should prioritize the use of wood-fibre insulation, and other nature-based material, in Europe and North America as these will deliver the maximum climate benefit, in the form of carbon storage.[3] We know our forests play a crucial role in storing carbon, our built environment could do the same. As architects dRMM have noted, "Given the urgency of the climate crisis we face, locking carbon into buildings now and in the short term will pay dividends long term. This outweighs the systems that involve emitting carbon upfront and seeking to find carbon storage technologies later. After all, humankind is yet to develop any carbon storage that works better than the tree".[4]

Chapter 4 introduced the new engineered timbers (Europe's term) or mass timbers (North America's term) which are enabling us to build big at height and at scale with timber. To date timber has been able to provide the structural frame for the majority of family homes in the USA, Canada, the Nordic countries, Scotland and millions more homes across Europe. However, it has *not* been able to compete with concrete and steel on bigger buildings of the type found in our town and city centres. That has changed thanks to advances in industrial engineered timber products and prefabrication techniques. Just as there was never any need to use anything other than timber to build the structure of a two-storey building, there is now no need to use anything other than timber to build one of 20 storeys. (Concrete foundations will still be needed). This means that wood's ability to substitute for carbon-intensive alternatives – especially concrete and steel – and safely to store carbon in the built environment has multiplied considerably. Therefore for the first time in 150 years we are looking at timber having the potential to be *the default primary structural construction material for the majority of buildings, small or large*, with all the ensuing climate benefits of substitution (CO_2 reduction) and storage (CO_2 removal), with the Potsdam Institute for Climate Impact Research reporting: ". . . a single-family house made of solid wood alone offsets the CO_2 emissions of 100 round-trip flights between Berlin and New York. This would transform building from a source of CO_2 to a powerful sink".[5]

However, this will only be possible if we have enough sustainable wood to meet a strong and steady growth in demand. This demand will ideally be augmented by the need for more wood to construct a rapidly growing number of timber-framed modular homes built in factories, but only if large-scale, off-site construction is developed in such a way that it can deliver quality, *affordable* homes which house buyers want to live in (Chapter 11).

To explore whether we can meet this imminent explosion in demand we examined the state of Europe's forests. We discovered in Chapter 7 that there is indeed a significant reserve of timber which, due to sustainable forest management, can meet our *current* level of demand in a sustainable way, including delivering good levels of biodiversity in our forests. As the IPCC have noted: "Sustainable forest management can maintain or enhance forest carbon stocks, and can maintain forest carbon sinks, including by transferring carbon to wood products, thus addressing the issue of sink saturation".[6]

It then became clear that if wood is to fulfil its maximum potential to *sequester*, *substitute* and *store* carbon we will need substantially more sustainable timber – and we will need it sooner rather than later. Where will this additional sustainable wood come from? From two main sources and one smaller one. Firstly, we should expand Europe's existing forest cover by up to 10 per cent (Chapter 9) with up to 3 per cent of this figure dedicated for nature either via natural forest regeneration (no human intervention needed) or specific planting for nature (intervention needed). Secondly, all farms should dedicate a minimum 10 per cent of their land to tree growing (Chapters 8 and 10). While up to half of the wood produced is likely to be used for energy more than a quarter could be used for the manufacture of wood products leaving the remaining quarter for the wider bioeconomy. It is *critical* that this "new" farm wood is grown in an agroforestry setting (Chapter 10) as this will enable farmers to maintain their current level of food production while at the same time increasing their incomes *and* the biodiversity on their farms. The third, smaller, source would comprise trees grown in urban settings.

Chapter 5 addressed concerns around timber buildings and the risk of fire in a way which, I hope, reassured anyone with worries on this subject. A timber-framed building, designed and built correctly, is just as safe as a building constructed from any other material.

Timber has a great deal to offer as we endeavour to find solutions to climate breakdown. What are the actions we need to take? What does it all add up to in the context of achieving net zero by 2050? Premised on two indisputable facts – that the substitution of carbon-intensive materials with nature-based materials, such as timber, *reduces* the amount of CO_2 entering the atmosphere, and that storing carbon in long-life timber products *removes* CO_2 from the atmosphere – I offer the following ten actions we must implement to maximize wood's potential to help save our world from climate breakdown.

1. Build timber-framed homes

New family homes (and all other new buildings up to three storeys) should be timber-framed. By *substituting* wood instead of the materials currently used we can both *avoid* carbon emissions and *store* carbon in the built environment. This change in approach, delivered at scale, will be highly beneficial for the climate: "Where wood carbon is transferred to harvested wood products, these can *store* carbon over the long-term and can *substitute* for emissions-intensive materials reducing emissions in other sectors" (emphasis added).[7]

CO_2 rating:
Reductions – high
Removals – high

2. Use engineered timber to build larger buildings

All tall and/or large new buildings should either be entirely constructed with an engineered-timber frame from the concrete foundations upwards, or they should be hybrid constructions with the lower third comprising concrete and steel and the upper two-thirds consisting of timber. By *substituting* the materials currently used – specifically concrete and steel – with wood we can both *avoid* carbon emissions and *store* carbon in the built environment. This change in approach, delivered at scale, will be highly beneficial for the climate: "Over 90% of concrete used in construction could be replaced with timber".[8]

CO_2 rating:
Reductions – high
Removals – high

3. Use nature-based, preferably wood-fibre, insulation

We should insulate all our existing buildings as close to a net-zero standard as practical and all new builds should meet the net-zero standard. This is the single biggest action we can take to decarbonize the built environment. We should do this using nature-based insulation materials, primarily wood-fibre, as this is the *only* way in which we can *also store* a significant amount of carbon in the built environment. In 20 years' time the majority of wood used in the built environment will be deployed for insulation if we are successful in optimizing timber's climate benefits. We will see a larger percentage of engineered timber being used for new builds in the Global South than in the North, but these buildings will also need to be insulated, mainly *against* the heat.

CO_2 rating:
Removals – very high

4. Expand existing forest cover
We should expand the existing European forest cover by up to 10 per cent, acknowledging that the expansion figure will vary from country to country. Forest expansion would be unlikely in Finland and Sweden given that their forest cover is already relatively high. However, it would be considerable in, say, the UK and the Republic of Ireland, involving at least a doubling of existing cover. This would increase the ability of Europe's forests to sequester and store up to an additional 10 per cent more carbon.

> We need more trees. We know that we need more for climate change, we know that we need more for biodiversity and we also need more for timber, we forget that. We talk a lot about food security but we import 81% of our timber that's eight and a half billion pounds. We need to have more homegrown timber and we need to use more in construction because it's better for the carbon footprint because if you are looking at bricks and steel there's clearly a big carbon footprint. Where we're looking at timber, we're locking up carbon for the long term so we need more of that.[9]

CO_2 rating:
Removals – high

5. Ensure 10 per cent tree cover on farms
We should introduce a mandatory minimum 10 per cent tree cover on all European farms with the trees planted in the most appropriate agroforestry configuration. This will *sequester* and *store* approximately an additional 8 per cent of CO_2 over and above the European forest's current level of sequestration.[10] In turn this additional planting will provide Europe with an onshore (not imported) supply of woody biomass which its remaining coal-fired power stations will need to use instead. This change,

combined with bioenergy with carbon capture and storage (BECCS), will help deliver a significant component of the necessary removals that both the UN's Intergovernmental Panel on Climate Change and the UK's independent Climate Change Committee envisage Europe and the wider world needing to become net zero by 2050.[11]

CO_2 rating:
Removals – very high

6. Boost urban tree cover

We should seek to maximize urban tree cover. The UK's urban environment currently has 16 per cent tree cover.[12] We should aim to reach the European average of 30 per cent.[13] This will *sequester* and *store* more carbon. Pockets of urban forest can contain as much carbon as tropical rainforests.[14] It will also bring the additional benefits for urban dwellers of cleaner air, the cooling impact of increased shade in the summer, noise reduction, the "softening" of hard landscapes, more biodiversity and added beauty. In London, where there are more trees than people, "It's estimated that 2,367,000 tonnes of carbon is stored in London's trees alone, with an estimated value of £147million to the UK's economy".[15]

CO_2 rating:
Removals – high
Health rating: high

7. Monetize wood's ability to store carbon and generate "gold standard" carbon offsets

We can generate carbon offsets (credits) against the carbon *stored* in the additional timber in the built environment and in the additional tree cover we create over and above what already exists.[16] In the case of the additional wood deployed in timber-framed homes and other smaller buildings these offsets could be used by the house-building companies themselves to help them achieve net-zero status for their own company, a journey many of them have now committed to.[17] (A house builder achieving net zero is of course a contribution to us all reaching net zero.) Those constructing the larger engineered-timber builds could sell their carbon offsets to help reduce the overall financial cost of the construction. In turn this would help make building with engineered timber more competitively priced against building in concrete and steel, and potentially cheaper if the price of carbon rises high enough. These carbon offsets entering the wider market could be bought by individuals and companies seeking to achieve net zero: "The recognition of durable wood products as an official carbon sink may create the opportunity to monetize carbon stored in buildings through carbon credit markets and create demand for timber construction".[18]

CO_2 rating:
Removals – will incentivize large-scale removals

8. Repair, reuse and recycle wood
The construction[19] and demolition of buildings provides the largest source of waste wood for recycling. Once we recognize the climate value of wood, we are much less likely to throw it away. By repairing, reusing and recycling wood products we can extend the period that the carbon is removed from the atmosphere. Recycling wood into chipboard (particleboard) is one of the most efficient ways of doing this. When there are no further uses for the wood, it should be burnt for energy. This ultimate use of wood for energy is the most efficient approach at the very end of the value chain: "Wood products from wood waste have carbon storage capabilities greater than the carbon discharged in the production process; thus, wood products offset the impact of global warming".[20]

CO_2 rating:
Removals – high

9. Make personal changes in our own lives

By familiarizing ourselves with the book's arguments we can make more informed purchasing decisions The most important action that most of us can take is to insulate our own homes to the highest possible standard (not just the loft but the walls and floors as well) using nature-based materials such as wood fibre: "Insulation has the greatest potential for reducing CO_2 emissions".[21]

CO_2 rating:

Reductions – individually small, collectively big
Removals – individually small, collectively big

10. Advocate

Spread the word. Make the case to all stakeholders, especially politicians (local, regional, national, supranational) and their staff. Praise their positive actions and criticize their poor actions by letter, email, telephone and via social media.

CO_2 rating:

Reductions – individually low, collectively high
Removals – individually low, collectively high

Taken together, these actions could reduce Europe's overall CO_2 emissions by at least 15 per cent, which is a conservative estimate. At a UK level these actions would have a bigger impact than ending *all* greenhouse gas emissions from the agriculture and international shipping sectors combined – two of the most difficult areas to reduce greenhouse gas emissions.[22] At an EU level these actions would be equivalent to ending *all* emissions from agriculture, international shipping and aviation.[23] It would result in Europe's built environment offsetting *all* of its emissions from *all* new build construction. It would also equate to Europe's built environment reducing the CO_2 emissions generated by its existing buildings (mainly through heating and cooling) by more than 50 per cent. It would mean we can shift construction from being a *source* of carbon emissions to becoming a carbon *sink*. The built environment would stop being part of the climate problem and instead become part of the climate *solution*.

And if that is not enough . . .

- The 10 per cent tree cover on all farms would take the majority of farms halfway to carbon neutrality.
- Biodiversity levels would increase across all European farmland, which covers 37 per cent of the continent's land mass.
- Urban tree planting would also increase biodiversity levels in our towns and cities which is where the majority of us live.
- Our built environment would become a healthier, more attractive place in which to live and work.
- The countries of Europe would take a large collective step towards implementing the legally binding UN Paris Agreement.

While writing this book, during 2022 and 2023, virtually every week saw a positive news story that highlighted wood's beneficial role in decarbonizing the built environment, which in turn was prompting fresh and innovative thinking around land use. There's not room to list them all, but I have picked out a few that showcase the most encouraging developments.

In the UK, the government set up the Timber in Construction Working Group, tasked with identifying significant actions which should be taken by the government, the construction industry and the timber industry in order to increase the number of timber and hybrid structures built in the UK. Their Timber in Construction Roadmap was published in December 2023 and its creation can be traced back to successful lobbying at the UN COP26 climate talks held in Glasgow.[24]

In the UK and Ireland, Aviva, a leading insurer, announced in 2023, "There are a growing number of developers looking to build more sustainably, both by using sustainable materials like engineered timber, and by adopting modern methods of construction. Aviva wants to embrace both: widening our underwriting appetite to insure commercial buildings using engineered timber, and using our risk management expertise to minimize associated risks".[25] Where Aviva lead others will follow, ensuring the construction of more large timber builds.

In the Netherlands, Amsterdam has launched a "Green Deal timber construction" in which more than 80 organizations are now involved, including the Dutch government, various municipalities, the transport authority, scientific institutions, housing associations, investors, architects, engineering firms and banks. As a result, from 2025, at least 20 per cent of all new buildings should be made of wood or other nature-based building materials.[26]

In Sweden, more than 20 per cent of mid-rise buildings are now timber framed (see Figure C.1).[27] In 2025 Swedish developers will embark on the "world's largest urban construction project in wood". "Stockholm Wood City" will be built in Sickla, an area in the south of the Swedish capital, with 7,000 office spaces and 2,000 homes, plus restaurants and shops across more than 250,000 square metres. It will take a decade to complete and will potentially set the stage for similar projects around the world.[28]

In Canada, in 2023 a report expounding the climate benefits of building in mass timber was released by the Climate Action Institute of the Royal Bank of Canada. The report noted that the "Widespread adoption of wood, specifically mass timber, as a substitute or complement to concrete and steel could cut embodied emissions in buildings by as much as 25%".[29]

The Council on Tall Buildings and Urban Habitat, based in Chicago, identified in 2022 that "there are now 139 mass timber buildings around the world of eight stories or higher, either complete, under construction or proposed".[30] Of those 71 per cent were in Europe, 18 per cent in North America, 10 per cent in Australia and 1 per cent in Asia. Worldwide there are currently 70 buildings of eight storeys or greater that are under construction or proposed. Many are in cities that already contain at least one tall timber building in countries with an existing tall timber economy but many are also new. It does appear that new height thresholds will go beyond the 25-storey (circa 90-metres high) record.

Figure C.1 Examples of Swedish mid-rise in Stockholm using CLT frames and clad in cedar tiles

Source: Folkhem.

In the USA it is estimated that the number of engineered/mass-timber buildings will double every year. If this could be replicated globally then the worldwide building industry could store more carbon than it emits by 2040.[31]

At COP28, held in December 2023 in the United Arab Emirates, a coalition of 17 countries made up of Australia, Canada, Congo, Costa Rica, Fiji, Finland, France, Germany, Ghana, Japan, Kenya, South Korea, Norway, Pakistan, Sweden, the UK and the USA, endorsed the following statement:

> Recognizing that wood from sustainably managed forests provides climate solutions within the construction sector, we commit to, by 2030, advancing policies and approaches that support low carbon construction and increase the use of wood from sustainably managed forests in the built environment. Such policies and approaches will result in reduced GHG emissions, and an increase in stored carbon.[32]

The idea of building and renovating more with wood is catching on. An architect designing a 30-storey engineered-timber skyscraper is increasingly unlikely to be considered a nutcase (to quote Mark Twain) and is more likely nowadays to be declared a climate superhero. Likewise, a farmer planting willow to make wood-fibre insulation is now viewed as a climate champion. Similarly, a joiner making modular homes in a factory for delivery to site is regarded not as an eccentric but as a climate defender. We built our way to climate breakdown. The time has come to build our way out – using wood. To quote John Schellnhuber once again,

> Without a radical shift in building, the Paris climate agreement will fail. However, if we replace steel and concrete with organic materials such as wood [. . .], we can avoid significant amounts of climate-damaging emissions. [. . .] With regenerative architecture, we could virtually build our way out of the climate crisis.[33]

And Austin Himes and Gwen Busby,

> We estimate that substituting conventional building materials with wood in half of new urban construction could provide 9% of global emissions reduction needed to meet 2030 targets for keeping global warming below 1.5°C. This could be a substantial piece of the emission reduction portfolio needed to meet the ambitious 2030 target without substantial changes to consumption, life-style or policy. Almost a tenth of necessary emissions reductions could come from choosing a different material for expected urban construction that will occur one way or the other.[34]

Wood is an amazing material. As a tree grows it *sequesters* and then *stores* carbon. When harvested the wood can go on *storing* carbon for decades in the form of timber in the built environment. Here it can *substitute* for carbon-intensive alternatives, most especially concrete.

Wood is nature's gift to humanity. If we use it wisely and sustainably it will undoubtedly help save us from climate breakdown.

Did Archbishop Trevor Huddleston live long enough to see Mandela freed, apartheid ended and a new South Africa born? He did indeed.

> Hope just means another world might be possible, not promised, not guaranteed. Hope calls for action; action is impossible without hope.[35]

Notes

Preface

1. Keir Starmer, Leader of the UK's Labour Party, played for the Law Department in the same league at the same time.

1 What is the problem we are trying to solve?

1. "2022 Global Status Report for Buildings and Construction", UN Environment Programme, https://www.unep.org/resources/publication/2022-global-status-report-buildings-and-construction.
2. N. Ruiz, L. Noe-Bustamante & N. Saber, "Coming of age", International Monetary Fund, March 2020, https://www.imf.org/en/Publications/fandd/issues/2020/03/infographic-global-population-trends-picture#:~:text=In%20just%2030%20years%2C%20the,to%20reach%20just%209.7%20billion.
3. United Nations, "Global issues: population", https://www.un.org/en/global-issues/population#:~:text=Africa%3A%20fastest%20growing%20continent,projected%20to%20double%20by%202050.
4. Extrapolated from Aureus Earth, "Today's challenge", https://aureusearth.com/about/.
5. Aureus Earth, "Today's challenge".
6. A. Cortese, "The embodied carbon conundrum: solving for all emission sources from the built environment", New Buildings Institute, 26 February 2020, https://newbuildings.org/embodied-carbon-conundrum-solving-for-all-emission-sources-from-the-built-environment/.
7. X. Zhong *et al.*, "Global greenhouse gas emissions from residential and commercial building materials and mitigation strategies to 2060", *Nature Communications* 12 (2021): 6126. doi.org/10.1038/s41467-021-26212-z.
8. Cortese, "The embodied carbon conundrum".
9. "Exajoule" is a measure of energy. Annual US energy use amounts to roughly 94 EJ. UN Environment Programme, "2022 Global Status Report for Buildings and Construction" (emphasis added).

10. We also need to switch to green energy but using green energy does not negate the need for better insulation because (1) we are still "short" of green energy, and (2) greener energy is no cheaper for the consumer to buy than fossil fuel energy.
11. B. Walker, "British homes among the worst insulated in Europe", *New Statesman*, 14 January 2022, https://www.newstatesman.com/chart-of-the-day/2022/01/british-homes-among-the-worst-insulated-in-europe.
12. R. Harrabin, "Climate change: UK government told to insulate more homes". BBC, 29 June 2022, https://www.bbc.co.uk/news/science-environment-61926275.
13. S. Yanatma, "Europe's energy crisis in data: which countries have the best and worst insulated homes?" *Euronews*, 9 December 2022, https://www.euronews.com/green/2022/12/09/europes-energy-crisis-in-data-which-countries-have-the-best-and-worst-insulated-homes.
14. Or by the time you have read this opening chapter.
15. J. Watts, "Concrete: the most destructive material on Earth". *The Guardian*, 25 February 2019, https://www.theguardian.com/cities/2019/feb/25/concrete-the-most-destructive-material-on-earth.
16. J. Timperley, "Q&A: Why cement emissions matter for climate change". Carbon Brief, 13 September 2018, https://www.carbonbrief.org/qa-why-cement-emissions-matter-for-climate-change.
17. I. Tiseo, "Carbon dioxide emissions in the European Union 1965–2022". Statista, 13 September 2023, https://www.statista.com/statistics/450017/co2-emissions-europe-eurasia/ and Timperley, "Why cement emissions matter for climate change".
18. Timperley, "Why cement emissions matter for climate change".
19. C. Hoffmann, M. Van Hoey & B. Zeumer, "Decarbonization challenge for steel", McKinsey, https://www.mckinsey.com/industries/metals-and-mining/our-insights/decarbonization-challenge-for-steel.
20. H. Ritchie, M. Roser & P. Rosado, "CO_2 and greenhouse gas emissions". Our World in Data, https://ourworldindata.org/co2-and-greenhouse-gas-emissions.
21. G. Churkina *et al.*, "Buildings as a global carbon sink", *Nature Sustainability* 3 (2020): 269–76, https://www.nature.com/articles/s41893-019-0462-4.
22. Watts, "Concrete: the most destructive material on Earth".
23. B. Gates, "A stunning statistic about China and concrete", GatesNotes, 25 June 2014, https://www.gatesnotes.com/Concrete-in-China.
24. World Population Review, "Egypt population", https://worldpopulationreview.com/countries/egypt-population.
25. M. Menshawy, "Why is Egypt building a new capital?" Al Jazeera, 5 July 2021,

https://www.aljazeera.com/opinions/2021/7/5/why-is-egypt-building-a-new-capital.

26. World Population Review, "World city populations", https://worldpopulationreview.com/world-cities.
27. World Population Review, "Rates of population growth around the world", https://worldpopulationreview.com/.
28. United Nations, "The human right to adequate housing", https://www.ohchr.org/en/special-procedures/sr-housing/human-right-adequate-housing.
29. B. King & C. Magwood, *Build Beyond Zero* (Washington, DC: Island Press, 2022), 81.
30. J. Tsui, "Plastic roads: an innovative solution to reduce waste", Kids.Earth.org, 15 November 2022, https://kids.earth.org/life-on-land/plastic-roads/.
31. "Why the built environment?" Architecture 2030, https://architecture2030.org/why-the-building-sector/#:~:text=Global%20building%20floor%20area%20is,every%20month%2C%20for%2040%20yea.
32. J. Pritchard, "Scrap to steel: the 5-step steel recycling process", Building Renewable, 17 October 2023, https://buildingrenewable.com/scrap-to-steel-steel-recycling-process/.
33. "Sustainable steel", Canadian Sheet Steel Building Institute, https://cssbi.ca/mid-rise-construction/sustainable-steel.
34. "Green Steel Tracker", Leadership Group for Industry Transition, https://www.industrytransition.org/green-steel-tracker/.
35. Swedish company Modvion built the world's first timber wind turbine tower in 2024, see https://modvion.com.
36. V. Viisainen & R. Bulleid, "A brighter future for UK steel". Green Alliance, 30 August 2023, https://green-alliance.org.uk/publication/a-brighter-future-for-uk-steel/.
37. *Ibid.*
38. S. Ellerbeck, "What is green steel and why does the world need more of it?" World Economic Forum, 11 July 2022, https://www.weforum.org/agenda/2022/07/green-steel-emissions-net-zero/.
39. S. Bhatnagar, "Five actions to improve the sustainability of steel", Ernst & Young Global, 20 July 2021, https://www.ey.com/en_gl/insights/energy-resources/five-actions-to-improve-the-sustainability-of-steel.
40. Ellerbeck, "What is green steel and why does the world need more of it?"
41. *Ibid.*
42. P. Griffin & G. Hammond, "The prospects for 'green steel' making in a net-zero economy: a UK perspective". *Global Transitions*, doi.org/10.1016/j.glt.2021.03.001.
43. J. Biggs, "Six low-carbon building materials for a greener construction

industry", Jobsite, 26 April 2021, https://www.procore.com/jobsite/six-low-carbon-building-materials-for-a-greener-construction-industry/#:~:text=A%20Global%20Alliance%20for%20Buildings,all%20global%20carbon%20dioxide%20emissions.

44. UN Environment Programme, "2022 Global Status Report for Buildings and Construction", 43.
45. H. Ritchie & M. Roser, "CO_2 emissions". Our World in Data, https://ourworldindata.org/co2-emissions.

2 How timber can decarbonize the built environment

1. Churkina *et al.*, "Buildings as a global carbon sink".
2. King & Magwood, *Build Beyond Zero*, 25.
3. G. Kilgore, "Carbon footprint of timber, wood, plywood (building materials footprint)", 8 Billion Trees, https://8billiontrees.com/carbon-offsets-credits/carbon-ecological-footprint-calculators/carbon-footprint-of-timber/.
4. A. Himes & G. Busby, "Wood buildings as a climate solution". *Developments in the Built Environment* 4 (2020). doi.org/10.1016/j.dibe.2020.100030.
5. "State of Europe's Forests 2020", Forest Europe, https://foresteurope.org/wp-content/uploads/2016/08/SoEF_2020.pdf.
6. The source for this estimate is a reputable expert in the field of timber construction.
7. M. Santoro, O. Cartus & J. Fransson, "Dynamics of the Swedish forest carbon pool between 2010 and 2015 estimated from satellite L-band SAR observations". *Remote Sensing of Environment* 270 (2022), doi.org/10.1016/j.rse.2021.112846.
8. J. Morison & R. Matthews, "Climate change and forest carbon. How much carbon is there in our woodlands?" Forest Researc, 23 May 2023, https://cdn.forestresearch.gov.uk/2023/07/Forest-carbon-factsheet.pdf.
9. Someone really does need to commission this piece of research as it is a gaping hole in our knowledge of the built environment.
10. R. Borrett, "Carbon jargon: how trees capture and store carbon". Carbon Neutral, April 2022, https://carbonneutral.com.au/carbon-jargon-how-trees-capture-and-store-carbon/.
11. Churkina *et al.*, "Buildings as a global carbon sink".
12. Innovawood, "Wood will play a tremendous role in the New European Bauhaus", Wood4Bauhaus, 4 June 2022, https://wood4bauhaus.eu/news/conference-report-w4b-featured-vatican-neb/.
13. U. Von der Leyen, "State of the Union Address 2020", European Commission. soteu_2020_en.pdf (europa.eu).
14. "Construction stored carbon 2021: a financial metric for carbon storage

in the built environment", ASN Bank & Climate Cleanup, https://climatecleanup.org/wp-content/uploads/2022/01/Construction-Stored-Carbon-V8-11-2021-5.pdf.

15. C. Magwood, T. Huynh & V. Olgyay, "The hidden climate impact of residential construction: zeroing in on embodied carbon emissions for low-rise residential buildings in the United States", Rocky Mountain Institute, Report 2023, https://rmi.org/insight/hidden-climate-impact-of-residential-construction/?utm_campaign=organic&utm_content=1680015757&utm_medium=social&utm_source=linkedin.
16. M. Childs, P. de Zylva & N. Rau, "Why we need more trees in the UK". Friends of the Earth, 22 June 2022, https://policy.friendsoftheearth.uk/download/why-we-need-more-trees-uk.
17. "Construction stored carbon 2021", 17; P. Leskinen *et al.*, "Substitution effects of wood-based products in climate change mitigation", European Forest Institute, 2018; R. Sathre & J. O'Connor, "Meta-analysis of greenhouse gas displacement factors of wood product substitution", *Environmental Science and Policy* 13:2 (2010); S. Rüter *et al.*, "ClimWood2030: Climate benefits of material substitution by forest biomass and harvested wood products: Perspective 2030 – Final", Thunen-Institut, 2016.
18. "Building sustainable cities with wooden skyscrapers", *The Economist*, 2 December 2021, https://www.economist.com/science-and-technology/2021/02/12/building-sustainable-cities-with-wooden-skyscrapers.
19. Churkina *et al.*, "Buildings as a global carbon sink". NB. kgC m^{-2} is the same as kgC divided by m^2, i.e, in this context "per" means "divide".
20. In Chapter 3 we shall explore how some wood-based building products can have a high level of embodied carbon.
21. UK Parliament, "The hammer-beam roof". https://www.parliament.uk/about/living-heritage/building/palace/westminsterhall/architecture/the-hammer-beam-roof-/.
22. Churkina *et al.*, "Buildings as a global carbon sink".
23. *Timber Construction: Barriers and Solutions*, Report for the All-Party Parliamentary Group for the Timber Industries, June 2023, https://timberdevelopment.uk/resources/timber-construction-barriers-and-solutions/ and United States Environmental Protection Agency, "Greenhouse Gas Equivalencies Calculator", https://www.epa.gov/energy/greenhouse-gas-equivalencies-calculator#results.
24. Churkina *et al.*, "Buildings as a global carbon sink".
25. Environmental Audit Committee, "Building to net zero: costing carbon in construction", House of Commons, 11 May 2022, https://committees.parliament.uk/publications/22427/documents/165446/default/.

26. Ramage in correspondence with the author.
27. N. Kurmayer, "EU to start measuring 'embodied' carbon emissions from buildings", Euractiv, 30 November 2021, https://www.euractiv.com/section/energy-environment/news/eu-to-start-measuring-embodied-carbon-emissions-from-buildings/.
28. "Biochar" is a charcoal produced from plant matter and stored in the soil as a means of removing CO_2 from the atmosphere.
29. Aureus Earth, https://www.aureusearth.com/.
30. Aureus Earth, "Aureus Earth and the University of Washington execute ground-breaking carbon offset transaction for a mass timber building", Cision PRWeb, 22 September 2022, https://www.prweb.com/releases/2022/9/prweb18906508.htm.
31. *Ibid.*
32. King & Magwood, *Build Beyond Zero*, 51.
33. *Ibid.*
34. L. Jensen, "EU Certification Framework for Carbon Removals", European Parliamentary Research Service, November 2023, https://www.europarl.europa.eu/RegData/etudes/ATAG/2023/754585/EPRS_ATA(2023)754585_EN.pdf.
35. J. Cummins, "Emission control: BloombergNEF reckons carbon offset prices could increase 50-fold by 2050", Stockhead, 21 January 2022, https://stockhead.com.au/energy/emission-control-bloombergnef-reckons-carbon-offset-prices-could-increase-50-fold-by-2050/.
36. J. Temple, "The growing signs of trouble for global carbon markets". *MIT Technology Review*, 2 November 2023, https://www.technologyreview.com/2023/11/02/1082765/the-growing-signs-of-trouble-for-global-carbon-markets/.
37. Childs, de Zylva & Rau, "Why we need more trees in the UK".

3 Insulation: the big climate win we have yet to deliver

1. "Wood fibre insulation: specification guidance", Woodknowledge Wales, https://woodknowledge.wales/category/resources/guidance/.
2. "Why invest in woodfibre insulation manufacturing in Wales", Woodknowledge Wales, https://woodknowledge.wales/why-invest-in-woodfibre-insulation-manufacturing-in-wales/.
3. D. Halpin, "'Substantial gap' between Government insulation target and delivery", *The Independent*, 26 June 2023, https://www.independent.co.uk/climate-change/news/uk-government-government-house-of-lords-england-b2363984.html; "Towards EU climate neutrality: progress, policy gaps and opportunities", https://climate-advisory-board.europa.eu/reports-

and-publications/towards-eu-climate-neutrality-progress-policy-gaps-and-opportunities.

4. "Final energy consumption in the residential sector by use, EU, 2021", Eurostat, https://ec.europa.eu/eurostat/statistics-explained/index.php?title=Energy_consumption_in_households#Energy_products_used_in_the_residential_sector.
5. *Ibid.*
6. "Nearly zero: a review of EU member state implementation of new build requirements", Buildings Performance Institute Europe, 22 June 2021, https://www.bpie.eu/publication/nearly-zero-a-review-of-eu-member-state-implementation-of-new-build-requirements/.
7. *Ibid.*
8. M. Weaver, "Brown pledges to build 'zero carbon' homes", *The Guardian*, 6 December 2006, https://www.theguardian.com/environment/2006/dec/06/politics.greenpolitics.
9. P. Oldfield, "UK scraps zero carbon homes plans", *The Guardian*, 10 July 2015, https://www.theguardian.com/environment/2015/jul/10/uk-scraps-zero-carbon-home-target.
10. S. Church, "'Altogether we are failing to deliver': The Lord Deben interview". The House, 10 June 2023, https://www.politicshome.com/thehouse/article/altogether-failing-deliver-chair-climate-committee-lord-deben-warns-dangers-climate-inaction.
11. "Why invest in woodfibre insulation", Woodknowledge Wales.
12. Climate Change Committee, "Progress towards reaching Net Zero in the UK", October 2023, https://www.theccc.org.uk/uk-action-on-climate-change/progress-snapshot/.
13. Climate Change Committee, "Independent Assessment: The UK's Heat and Buildings Strategy", March 2022, https://www.theccc.org.uk/publication/independent-assessment-the-uks-heat-and-buildings-strategy/.
14. B. Wilson, "Armitt: exit strategy from fossil fuels will boost energy security and lower household costs", National Infrastructure Commission, 29 September 2022, https://nic.org.uk/insights/armitt-exit-strategy-from-fossil-fuels-will-boost-energy-security-and-lower-household-costs/.
15. M. Savage, "Cameron's decision to cut 'green crap' now costs each household in England £150 a year". *The Guardian*, 19 March 2022, https://www.theguardian.com/money/2022/mar/19/david-cameron-green-crap-energy-prices.
16. A. Esser *et al.*, "Comprehensive study of building energy renovation activities and the uptake of nearly zero-energy buildings in the EU", European

Commission report, November 2019, https://energy.ec.europa.eu/system/files/2019-12/1.final_report_0.pdf.

17. R. Boydell, "Five numbers that lay bare the mammoth effort needed to insulate Britain's homes". The Conversation, 14 September 2021, https://theconversation.com/five-numbers-that-lay-bare-the-mammoth-effort-needed-to-insulate-britains-homes-162540.
18. "Charity warns UK Government could miss its legal fuel poverty commitments by over 150 years", National Energy Action, 5 December 2022, https://www.nea.org.uk/news/charity-warns-uk-government-could-miss-its-legal-fuel-poverty-commitments-by-over-150-years/.
19. C. Pavel & D. Blagoeva, "Competitive landscape of the EU's insulation materials industry for energy-efficient buildings, revised edition", Publications Office of the European Union, https://op.europa.eu/en/publication-detail/-/publication/23dd9566-0289-11e9-adde-01aa75ed71a1/language-en.
20. *Ibid.*
21. "Which insulation most effectively reduces whole-life carbon (embodied and operational)?" Mantle Developments, 26 August 2022, https://mantledev.com/insights/reduces-whole-life-carbon/#:~:text=The%20results%20showed%20that%3A,higher%20than%20the%20other%20materials.
22. "Environmental Product Declaration (EPD): overview", Ecochain, https://ecochain.com/blog/environmental-product-declaration-epd-overview/.
23. C. Hill, A. Norton & J. Dibdiakova, "A comparison of the environmental impacts of different categories of insulation materials", *Energy and Buildings* 162 (2018), https://www.sciencedirect.com/science/article/abs/pii/S0378778817334679.
24. *Ibid.*
25. "Embodied carbon values in common insulation materials", KPMB Architects, May 2021, https://www.kpmb.com/wp-content/uploads/2021/04/KPMB-LAB_Embodied-Carbon-in-Insulation.pdf.
26. J. Bull, "Embodied carbon in insulation", GreenSpec, https://www.greenspec.co.uk/building-design/embodied-carbon-of-insulation/.
27. Some cement manufacturers are considering switching to biomass as their energy source to drive down their levels of embodied carbon.
28. "Flumroc will melt rock with renewable energy in future", ImmoInvest, March 2023, https://www.immo-invest.ch/en/flumroc-schmilzt-gestein-kuenftig-mit-erneuerbarer-energie/?utm_source=rss&utm_medium=rss&utm_campaign=flumroc-schmilzt-gestein-kuenftig-mit-erneuerbarer-energie.
29. Lynn in conversation with the author.

30. K. Diab, "CRCF: The EU's carbon certification removal failure", Carbon Market Watch, 20 February 2024, https://carbonmarketwatch.org/2024/02/20/crcf-the-eus-carbon-removal-certification-failure/.
31. Readers may wonder why cork is not mentioned here. While cork is an excellent insulating material its sustainable availability is severely restricted compared to wood fibre. In Europe only Portugal is a major producer, with smaller amounts available in Spain, France and Italy, whereas wood fibre is widely and sustainably available across Europe.
32. "Wood4Bauhaus Bioeconomy Side Event", European Panel Federation, https://europanels.org/press-release-building-the-bioeconomy-with-nature-based-materials-and-solutions-within-the-new-european-bauhaus/.
33. Churkina *et al.*, "Buildings as a global carbon sink".
34. "New production line for wood fibre insulation materials", Sonae Arauco, https://www.sonaearauco.com/news/sonae-arauco-invests-in-the-future-of-its-agepan-system-for-construction-in-meppen-site/.
35. "Why invest in woodfibre insulation", Woodknowledge Wales.

4 Wooden skyscrapers

1. P. Van der Lugt & A. Harsta, *Tomorrow's Timber: Towards the Next Building Revolution* (Naarden, NL: Material District 2020), 80.
2. "Finland's tallest wooden apartment building rises to crest height", Tekniikan Maailma, 15 March 2019, https://tekniikanmaailma.fi/suomen-korkein-puukerrostalo-nousee-harjakorkeuteen-talta-14-kerroksinen-talo-nayttaa-nyt/.
3. "Adhesives in engineered timber", New Nordic Timber, https://newnordictimber.com/articles/adhesives-in-engineered-timber.
4. "About Mjøstårnet", Moelven, https://www.moelven.com/mjostarnet/.
5. Van der Lugt & Harsta, *Tomorrow's Timber*, 119.
6. "A sustainable intersection of old and new", Tengbom, https://en.tengbom.se/project/trikafabriken-2/.
7. "Coroner issues report on preventing future air pollution deaths", Client Earth, 21 April 2021, https://www.clientearth.org/latest/press-office/press/coroner-issues-report-on-preventing-future-air-pollution-deaths-clientearth-reaction/#:~:text=In%20December%202020%2C%20the%20Southwark,Deaths%20report%20following%20the%20inquest.
8. K. Taylor-Smith, "The threat to public health from tyre pollution could become major issue for regulators", Chartered Institute for Environmental Health, June 2022, https://www.cieh.org/ehn/environmental-protection/2022/june/the-threat-to-public-health-from-tyre-pollution-could-become-

major-issue-for-regulators/#:~:text=Tyres%20are%20made%20from%20 synthetic,the%20air%2C%20soil%20and%20water.

9. "Our position on improving air quality", Sustrans, 9 April 2019, https://www.sustrans.org.uk/our-blog/policy-positions/all/all/our-position-on-improving-air-quality/.
10. O. Wainwright, "'I want to caress the lift!': the eco office block miracle made entirely from wood", *The Guardian*, 30 January 2023, https://www.theguardian.com/artanddesign/2023/jan/30/caress-lift-eco-office-block-miracle-wood-timber-black-white.

5 Fire: timber buildings and safety

1. R. O'Brocki, "What the Fire Service needs to know about tall wood buildings". Firehouse, 11 June 2018, https://www.firehouse.com/operations-training/article/21008544/what-the-fire-service-needs-to-know-about-tall-wood-buildings.
2. N. Barker, "Are mass-timber buildings a fire safety risk?" Dezeen, 22 March 2023, https://www.dezeen.com/2023/03/22/mass-timber-fire-safety-timber-revolution/.
3. "Bessie Surtees House", Historic England, https://historicengland.org.uk/get-involved/visit/bessie-surtees-house/.
4. Typical dimensions of timber used in historic post-and-beam structures range between 10–30 cm or larger, whereas modern timber-frame dimensions are smaller, ranging mainly between 4–15 cm and for a few roofing roles up to 22 cm.
5. Department for Environment, Food and Rural Affairs, "Timber in construction roadmap", Policy Paper, 28 December 2023, https://www.gov.uk/government/publications/timber-in-construction-roadmap/timber-in-construction-roadmap.
6. "Timber around the world: England v Scotland", Structural Timber Association, 3 November 2021, https://www.structuraltimber.co.uk/news/timber-around-the-world-england-v-scotland/#:~:text=Historically%2C%20structures%20would%20be%20built%20using%20materials%20that,and%20clay%20were%20the%20more%20easily%20obtainable%20materials.
7. Sara Culture House, https://whitearkitekter.com/project/sara-cultural-centre/
8. Structural Timber Association, https://www.structuraltimber.co.uk/.
9. Timber Development UK, https://timberdevelopment.uk/.
10. Some worry that FTAs and the glues that hold engineered timbers together may release noxious fumes in a fire. The reality is that the heat and smoke

from the burning contents will have already killed any occupant that has failed to evacuate before the timber and its FTA coating catches fire.

11. "Flame retardant treatments for timber: an overview", Timber Development UK, 19 January 2023, https://timberdevelopment.uk/resources/flame-retardant-treatments-for-timber-an-overview/.
12. "Crosslam Timber/CLT – fire resistance and rating", Greenspec, https://www.greenspec.co.uk/building-design/crosslam-timber-fire-resistance-and-rating/#:~:text=CLT's%20fire%20resistance%20is%20provided%20through%20'charring'.%20As,strength%20and%20becomes%20a%20black%20layer%20of%20'char'.
13. "Timber typologies: understanding options for timber construction", Timber Development UK, 7 June 2023, https://timberdevelopment.uk/resources/timber-typologies/.
14. Milner in conversation with the author.
15. "Fire resistant building", Swedish Wood, 2014, https://www.swedishwood.com/publications/wood-magazine/2014-2/fire_resistant_building/#:~:text=%E2%80%9CThe%20fact%20is%20that%20wood%20has%20many%20fire,for%20a%20long%20time%2C%20even%20during%20a%20fire.%E2%80%9D.
16. Bronto Skylift, https://brontoskylift.com/wp-content/uploads/2019/04/S_XR-range-US.pdf.
17. P. Apps, *Show me the Bodies: How we let Grenfell Happen* (London: Oneworld, 2022).
18. Swedish Wood, Timber Development UK & Structural Timber Association, "Fire safety wood in construction", https://timberfiresafety.org/.
19. "How the world's tallest timber building will withstand fire", Green Building, 20 June 2018, http://greenbuilding.co.uk/how-the-worlds-tallest-timber-building-will-withstand-fire/.
20. "Insuring mass timber: solving the Catch-22 conundrum", Built by Nature, 6 April 2022, https://builtbn.org/news/insuring-mass-timber-solving-the-catch-22-conundrum/179.
21. *Ibid.*
22. *Ibid.*
23. *Ibid.*
24. Built by Nature, https://builtbn.org/about.
25. "Eight reasons to use wood fibre insulation", Back to Earth, https://www.backtoearth.co.uk/news/eight-reasons-to-use-wood-fibre-insulation/.

6 Forests: an important asset in tackling climate breakdown

1. H. Mauser, "How have forest resources in the European Union developed?" European Forest Institute, https://efi.int/forestquestions/q15#:~:text=In%20the%20EU%2C%20the%20forest,of%20the%20EU%20territory%20(approx.
2. Food and Agriculture Organization of the United Nations, "The state of the world's forests 2020". https://www.fao.org/state-of-forests/en/.
3. "State of Europe's Forests 2020", Forest Europe.
4. "Woodland statistics", Forest Research, https://www.forestresearch.gov.uk/tools-and-resources/statistics/statistics-by-topic/woodland-statistics/.
5. World Bank, "Forest Area (% of land): European Union", https://data.worldbank.org/indicator/AG.LND.FRST.ZS?locations=EU.
6. Deciduous trees shed their leaves in the winter. Most broadleaf trees are deciduous.
7. Needle-shaped leaves that stay on the tree all year round, except for the larch (the only deciduous conifer in Europe) which sheds its needles in the autumn.
8. Mauser in correspondence with the author.
9. A. Kilpeläinen & H. Peltola, "Carbon sequestration and storage in European forests" in L. Hetemäki, J. Kangas & H. Peltola (eds), *Forest Bioeconomy and Climate Change: Managing Forest Ecosystems* (Cham: Springer, 2022).
10. R. Giffen *et al.*, "Storing more carbon by improving forest management in the Acadian Forest of New England, USA", MDPI, https://www.mdpi.com/1999-4907/13/12/2031.
11. Joint Nature Conservation Committee, "Area of forestry land certified as sustainably managed". https://jncc.gov.uk/our-work/ukbi-b1b-sustainable-forestry/#:~:text=Downloads-,Key%20results,%25%20or%2044%25%20since%202007.
12. "Bringing woodland into management: the missed opportunities in England and Wales", Royal Forestry Society, https://rfs.org.uk/wp-content/uploads/2021/03/woodland-management-missed-opportunities.pdf.
13. Mauser, "How have forest resources in the European Union developed?"
14. "Deforestation", Greenpeace, https://www.greenpeace.org.uk/challenges/forests/deforestation/?gclid=EAIaIQobChMIidKqs7Pv_gIV4ejtCh3aEQCHEAAYAiAAEgKl3vD_BwE.
15. In the UK private sector planting usually takes place within two years. In the public sector it is not uncommon to wait 4–5 years to avoid applying spot chemical treatment for weevils. The public sector doesn't face the same financial pressures so will sacrifice income in order to reduce chemical use.

16. Deep peat being greater than 50 cm thick. Planting trees on peat that is less than 50 cm thick is a carbon positive activity.
17. P. Ford, "The bog blog: everything you never knew you needed to know about peat". Council for the Protection of Rural England, 28 July 2021, https://www.cpre.org.uk/explainer/the-bog-blog-everything-you-never-knew-you-needed-to-know-about-peat/.
18. Berg in conversation with the author.
19. "Ancient woodland", Woodland Trust, https://www.woodlandtrust.org.uk/trees-woods-and-wildlife/habitats/ancient-woodland/.
20. E. Harris, "Biodiversity, forestry and wood: an analysis of the biodiversity benefits of modern forestry and wood production", Confor report, https://www.confor.org.uk/media/247794/confor-biodiversity-forestry-report.pdf.
21. "Woodland statistics", Forest Research, https://www.forestresearch.gov.uk/tools-and-resources/statistics/statistics-by-topic/woodland-statistics/.
22. Climate Change Committee, "Land Use: Policies for a Net Zero UK", January 2020, https://www.theccc.org.uk/publication/land-use-policies-for-a-net-zero-uk/.
23. "Woodland statistics: UK tree planting still 'way off target'", Confor, 15 June 2023, https://www.confor.org.uk/news/latest-news/woodland-statistics-uk-tree-planting-still-way-off-target/.
24. Heartwood Forest, https://heartwood.woodlandtrust.org.uk and the Northern Forest, https://www.woodlandtrust.org.uk/about-us/what-we-do/we-plant-trees/the-northern-forest/.
25. For comparison, a football pitch is 0.6–0.8 hectares.
26. Andy Howard, director of Pennine Forestry who manage the project, in conversation with the author.
27. "Our founder", the Heart of England Forest, https://heartofenglandforest.org/our-founder.
28. J. Aldred, "Felix Dennis plants his millionth tree". *The Guardian*, 20 September 2013, https://www.theguardian.com/environment/2013/sep/20/felix-dennis-million-trees.
29. D. Jackson, "The felled Sycamore Gap tree leaves a hole in Northumbrians' hearts". UnHerd, 29 September 2023, https://unherd.com/thepost/the-felled-tree-leaves-a-gap-in-northumbrians-hearts/.
30. "Common non-native trees", Woodland Trust, https://www.woodlandtrust.org.uk/trees-woods-and-wildlife/british-trees/common-non-native-trees/.
31. J. Haugh, "UK falls to become world's third-largest net importer of timber", *Forestry Journal*, 3 October 2023, https://www.forestryjournal.co.uk/news/23829587.uk-falls-become-worlds-third-largest-net-importer-timber/.
32. "Forestry and timber Q&A", Confor, https://www.confor.org.uk/resources/

publications/confor-publications/forestry-and-timber-qa/#:~:text=The%20 UK%20already%20imports%2080,decking%2C%20packaging%20and%20 much%20more.

33. Department for Environment, Food and Rural Affairs, "United Kingdom Food Security Report 2021: Theme 2: UK food supply sources", https://www.gov.uk/government/statistics/united-kingdom-food-security-report-2021/united-kingdom-food-security-report-2021-theme-2-uk-food-supply-sources#:~:text=Headline,%C2%A321.4%20billion%20was%20exported.
34. "Confor accuses UK Government of 'total policy failure' on tree planting", Confor, 16 June 2022, https://www.confor.org.uk/news/latest-news/confor-accuses-uk-government-of-total-policy-failure-on-tree-planting/.
35. E. Ares, S. Coe & E. Uberoi, "Tree Planting in the UK". House of Commons Library Briefing Paper Number 9084, 2 June 2021, https://researchbriefings.files.parliament.uk/documents/CBP-9084/CBP-9084.pdf.
36. *Ibid.*
37. The period of time between planting and felling.
38. I. Davies, "Sustainable construction timber: sourcing and specifying local timber", Forestry Commission Research Report, https://www.forestresearch.gov.uk/publications/sustainable-construction-timber/ and D. Ridley-Ellis, D. Gil-Moreno & A. Harte, "Strength grading of timber in the UK and Ireland in 2021", *International Wood Products Journal* 13:2 (2022): 127–36.
39. T. Barnes & A. Leitch, "The National Wood Strategy: a path towards a sustainable forest products industry in England". https://www.nationalwoodstrategy.co.uk/.
40. *Ibid.*
41. M. Wilkinson, G. Xenakis & J. Morison, "The carbon balances of two contrasting forest stands growing in the UK", Forest Research, September 2023, https://cdn.forestresearch.gov.uk/2023/09/Research-note_The-carbon-balances-of-two-forest-stands.pdf.
42. Harris, "Biodiversity, forestry and wood".

7 Is there enough sustainable wood?

1. Environmental Audit Committee, "Seeing the wood for the trees: the contribution of the forestry and timber sectors to biodiversity and net zero goals: report summary", 19 July 2023, https://publications.parliament.uk/pa/cm5803/cmselect/cmenvaud/637/summary.html.
2. Forest Stewardship Council, https://fsc.org/en and Programme for the Endorsement of Forest Certification, https://www.pefc.org/.
3. "Things you may not know about forests in Europe: forest area, annual growth and fellings, but also protected areas all continue to increase", Finnish

Forest Association, https://forest.fi/article/things-you-may-not-know-about-forests-in-europe-forest-area-annual-growth-and-fellings-but-also-protected-areas-all-continue-to-increase/#49ddbae9.

4. Subsequently revised to 63 per cent as per the table.
5. Eurostat, "Forests, forestry and logging", https://ec.europa.eu/eurostat/statistics-explained/index.php?title=Forests,_forestry_and_logging#Forests_in_the_EU.
6. "State of Europe's Forests 2020", 90.
7. Kalle, K. *et al.*, "Overview of global long-distance road transportation of industrial roundwood", *Croatian Journal of Forest Engineering* 45:1 (2017), https://crojfe.com/archive/volume-45-no.1/overview-of-global-long-distance-road-transportation-of-industrial-roundwood/.
8. "Pan-European indicators of sustainable forest management", Forest Research, https://www.forestresearch.gov.uk/tools-and-resources/statistics/statistics-by-topic/international-returns/forest-europe/.
9. M. Vizzarri *et al.*, "Setting the forest reference levels in the European Union: overview and challenges", *Carbon Balance and Management* 16:1 (2021), https://www.researchgate.net/publication/353611493_Setting_the_forest_reference_levels_in_the_European_Union_overview_and_challenges.
10. J. Mitchell, "Net zero could drive up the global demand for timber, putting at risk the world's forests", Investment Monitor, 9 November 2022, https://www.investmentmonitor.ai/features/net-zero-drive-up-global-demand-timber-forests/.
11. European Commission, "A sustainable bioeconomy for Europe: strengthening the connection between economy, society and the environment – Updated bioeconomy strategy", https://op.europa.eu/en/publication-detail/-/publication/edace3e3-e189-11e8-b690-01aa75ed71a1/language-en.
12. P. Patel, "For sustainable chemicals, replace petroleum with wood", *Anthropocene Magazine and Future Earth*, 20 February 2020, https://www.anthropocenemagazine.org/2020/02/for-sustainable-chemicals-replace-petroleum-with-wood/#:~:text=Plastics%2C%20detergents%2C%20paints%2C%20medicines,at%20the%20University%20of%20Leuven.
13. European Commission, "A sustainable bioeconomy for Europe", 5.
14. Cited in the President's Council of Advisors on Science and Technology, "Report to the President: Biomanufacturing to advance the bioeconomy", December 2022, 12; https://www.whitehouse.gov/wp-content/uploads/2022/12/PCAST_Biomanufacturing-Report_Dec2022.pdf.
15. European Commission, "A sustainable bioeconomy for Europe", 33.

16. "Russia's Boreal forests", World Wildlife Fund, https://wwfeu.awsassets.panda.org/downloads/russia_forest_cc_final_13nov07.pdf.
17. K. Press-Kristensen & M. Tolotto, "Where there's fire, there's smoke: emissions from domestic heating with wood", European Environmental Bureau and Green Transition Denmark, September 2021, https://eeb.org/wp-content/uploads/2021/09/Where-theres-fire-theres-smoke_domestic-heating-study_2021.pdf.
18. A recent study (July 2022) conducted by CERIC Laboratoires in France revealed that "between 2010 and 2020, emissions of the smallest particles have been reduced by 45 per cent in the residential and tertiary sectors, largely thanks to technical developments in wood-burning appliances and the replacement of ageing equipment", "Impact de la qualité du combustible bois et de l'évolution du par d'appareils sur la qualité de l'air", https://www.laboratoire-ceric.com.
19. DEFRA, "Emissions of air pollutants in the UK – particulate matter (PM10 and PM2.5)", https://www.gov.uk/government/statistics/emissions-of-air-pollutants/emissions-of-air-pollutants-in-the-uk-particulate-matter-pm10-and-pm25.
20. T. Lecomte *et al.*, "Best available techniques (BAT) reference document for large combustion plants", European Union, Joint Research Centre, https://op.europa.eu/en/publication-detail/-/publication/c31e5e11-db60-11e7-a506-01aa75ed71a1/language-en.
21. European state forest organizations do not sell hardwood timber to China.
22. "Forest products in the global bioeconomy: enabling substitution by wood-based products and contributing to the Sustainable Development Goals", European Forest Institute, 15 November 2021, https://efi.int/publications/forest-products-global-bioeconomy-enabling-substitution-wood-based-products-and.
23. Food and Agriculture Organization of the United Nations, "Forestry production and trade", https://www.fao.org/faostat/en/#data/FO.
24. G. Moreau *et al.*, "Opportunities and limitations of thinning to increase resistance and resilience of trees and forests to global change", *Forestry: An International Journal of Forest Research* 95:5 (2022): 595–615.
25. G. Kerr & J. Haufe, "Thinning practice: a silvicultural guide", Forestry Commission, January 2011, https://cdn.forestresearch.gov.uk/2011/01/silviculture_thinning_guide_v1_jan2011.pdf.
26. A. Węgiel *et al.*, "Effect of logging residue removal and mechanical site preparation on productivity of the subsequent Scots pine (*Pinus sylvestris L.*) stands", *Annals of Forest Science* 80:5 (2023).
27. F. Aldebei & M. Dombi, "Mining the built environment: telling the story of urban mining", *Buildings* 11:9 (2021): 388.

28. Y. Niu *et al.*, "Prolonging life cycles of construction materials and combating climate change by cascading: the case of reusing timber in Finland", *Resources, Conservation and Recycling* 170 (2021), https://www.sciencedirect.com/science/article/pii/S0921344921001622?via%3Dihub.
29. G. Spocchia, "City of London demands alternatives to demolition for new schemes". *Architects' Journal*, 9 March 2023, https://www.architectsjournal.co.uk/news/city-of-london-demands-alternatives-to-demolition-for-new-schemes.
30. European Panel Federation, https://europanels.org/.
31. "About waste wood", Wood Recyclers Association, https://woodrecyclers.org/about-waste-wood/#:~:text=Uses%20for%20recycled%20waste%20wood,play%20areas%20and%20filter%20beds.
32. Rilegno, https://www.rilegno.org/.
33. "Assessing the carbon-related impacts and benefits of timber in construction products and buildings", Timber Development UK, November 2021, https://timberdevelopment.uk/resources/assessing-the-carbon-related-impacts-and-benefits-of-timber-in-construction/ and "About waste wood", Wood Recyclers' Association.
34. *Ibid.*
35. Z. Stein, "Life cycle assessment (LCA)", Carbon Collective, https://www.carboncollective.co/sustainable-investing/life-cycle-assessment-lca.
36. "Treelogy, Part II: the tree and timber", dRMM, 22 November 2021, https://drmmstudio.com/article/treelogy-part-ii-the-tree-and-timber/.
37. "A pioneering achievement", Timber Development UK, 27 November 2023, https://timberdevelopment.uk/a-pioneering-achievement-timber-office-complex-wins-2-prizes/#:~:text=The%20Black%20%26%20White%20Building%20is,the%20Commercial%20and%20Sustainability%20categories and T. Alouani-Roby, "What's black and white and timber all over?" IndesignLive, 16 February 2023, https://www.indesignlive.com/projects/waugh-thistelton-timber#:~:text=Notably%2C%20the%20building%20is%20also,rather%20than%20the%20purely%20decorative.
38. L. Alter, "London's 'black and white' building is complete – and it's a knockout", Treehugger, 20 January 2023, https://www.treehugger.com/black-and-white-building-is-completed-and-it-s-a-knockout-7096824#:~:text=It's%20designed%20for%20disassembly%2C%20%22screwed,to%20be%20truly%20considered%20sustainable.&text=Others%20would%20argue%20that%20this%20isn't%20how%20the%20math%20works.
39. "Wood industry aims to double the amount of timber used annually in the European built environment by 2050", European Confederation of Woodworking Industries, press release, 26 November 2021,

https://www.cei-bois.org/_files/ugd/5b1bdc_9cd0c04bfc084a788ebadfe3f986087a.pdf.

40. "Paper packaging", FERN, https://www.fern.org/issues/paper-packaging/.

8 Increasing wood supply

1. "What is bioenergy?", Bioenergy Europe, https://bioenergyeurope.org/about-bioenergy.html.
2. "Biomass supply statistical report", Bioenergy Europe, https://bioenergyeurope.org/statistical-report.html.
3. Watkins in correspondence with the author.
4. M. Christian, "Drax: the largest biomass power station in the world", AquaSwitch, https://www.aquaswitch.co.uk/blog/drax-biomass-power-station/
5. L. Fernandez, "Biomass plant installed capacity in the United Kingdom (UK) 2020, by site", Statista, https://www.statista.com/statistics/896770/biomass-plant-installed-capacity-united-kingdom-uk/.
6. "UK: Joint NGO statement on BECCS", Land & Climate Review, 23 March 2021, https://www.landclimate.org/uk-joint-ngo-statement-on-beccs/.
7. P. Fennell, "Making the case for carbon capture and storage", *The Guardian*, 3 April 2023, https://www.theguardian.com/environment/2023/apr/03/making-the-case-for-carbon-capture-and-storage.
8. B. Robertson, "Carbon capture remains a risky investment for achieving decarbonization", Institute for Energy Economics and Financial Analysis, https://ieefa.org/resources/carbon-capture-remains-risky-investment-achieving-decarbonisation?gclid=EAIaIQobChMI0cmmjZma_QIVW-ztCh23rwHpEAAYAiAAEgKc_PD_BwE.
9. J. Beddington *et al.*, "Letter from scientists to the EU Parliament regarding forest biomass", 9 January 2018, scientist-letter-on-eu-forest-biomass-796-signatories-as-of-january-16-2018.pdf (wordpress.com).
10. R. Irslinger *et al.*, "Over 550 scientists back bioenergy as part of 'Climate Smart' forest management", Association for Renewable Energy & Clean Technology, 31 October 2022, https://www.r-e-a.net/over-550-scientists-back-bioenergy-as-part-of-climate-smart-forest-management/.
11. Climate Change Committee, "Progress in reducing emissions: 2023 Report to Parliament", 248; https://www.theccc.org.uk/wp-content/uploads/2023/06/Progress-in-reducing-UK-emissions-2023-Report-to-Parliament-1.pdf.
12. H. Hassi, "Introduction to forest-based bioeconomy", Forestbiofacts, https://forestbiofacts.com/introduction-to-forest-based-bioeconomy/.
13. "What is Insulate Britain and what does it want?" BBC News, 17 November 2021, https://www.bbc.co.uk/news/uk-58916326.

14. European Commission, "Renovation wave: doubling the renovation rate to cut emissions, boost recovery and reduce energy poverty", press release, 14 October 2020, https://ec.europa.eu/commission/presscorner/detail/en/IP_20_1835.
15. European Confederation of Woodworking Industries, https://www.cei-bois.org.
16. A. Kallio, "Wood-based textile fibre market as part of the global forest-based bioeconomy", *Forest Policy and Economics* 123 (2021).
17. L. Hickman, "Timeline: how BECCS became climate change's 'saviour' technology", Carbon Brief, 13 April 2016, https://www.carbonbrief.org/beccs-the-story-of-climate-changes-saviour-technology/.
18. R. Weyler, "The great carbon capture scam", Greenpeace, https://www.greenpeace.org/international/story/54079/great-carbon-capture-scam/ and FERN, "Six problems with BECCS", https://www.fern.org/publications-insight/six-problems-with-beccs-57/?gclid=EAIaIQobChMI2JXbm6Ga_QIVjsvtCh151gAcEAAYAiAAEgKsVfD_BwE.
19. "UK: Joint NGO Statement on BECCS".
20. B. Metz *et al.*, "IPCC Special Report on Carbon Dioxide Capture and Storage", IPCC 2005, https://www.ipcc.ch/site/assets/uploads/2018/03/srccs_wholereport-1.pdf.

9 Is there enough land?

1. C. Alcantara *et al.*, "Mapping the extent of abandoned farmland in Central and Eastern Europe using MODIS time series satellite data", *Environmental Research Letters* 8:3 (2013).
2. Office for National Statistics, "National population projections: 2020-based interim", https://www.ons.gov.uk/peoplepopulationandcommunity/populationandmigration/populationprojections/bulletins/nationalpopulationprojections/2020basedinterim.
3. G. Daskalova & J. Kamp, "Abandoning land transforms biodiversity", *Science* 380:6645 (2023), https://www.science.org/doi/10.1126/science.adf1099.
4. It should, however, be noted that there were agricultural practices in the old USSR that were unquestionably beneficial, for instance, the planting of trees to act as agricultural shelter belts as still found today in both Ukraine and Russia; see O. Petrovich, "Ecosystem services providing shelterbelts", https://core.ac.uk/download/pdf/14063306.pdf and S. Schwarz, "USSR: the Shelterbelt program in its relation to other drought control projects", *Land Economics* 25:4 (1949): 360–62.
5. "Farms and farmland in the European Union: statistics", Eurostat, https://ec.europa.eu/eurostat/statistics-explained/index.php?title=Farms_and_farmland_in_the_European_Union_-_statistics#Farms_in_2020.

6. "Agriculture in Romania", Flanders Investment & Trade Market Survey 2017, https://www.flandersinvestmentandtrade.com/export/sites/trade/files/market_studies/2017_Agriculture_Romania.pdf.
7. Alcantara, "Mapping the extent of abandoned farmland".
8. K. Valujeva *et al.*, "Abandoned farmland: past failures or future opportunities for Europe's Green Deal? A Baltic case-study". *Environmental Science & Policy* 128 (2022): 175–84.
9. *Ibid.*
10. "Agricultural land abandonment in the EU within 2015–2030". Joint Research Centre, Policy Insights 2018, https://joint-research-centre.ec.europa.eu/system/files/2018-12/jrc113718.pdf and R. Barranco *et al.*, "Territorial facts and trends in the EU rural areas within 2015–2030", https://op.europa.eu/en/publication-detail/-/publication/e67b137a-fcf6-11e8-a96d-01aa75ed71a1/language-en.
11. R. Conniff *et al.*, "Could abandoned agricultural lands help save the planet?" Yale Environment 360, 10 December 2019, https://e360.yale.edu/features/could-abandoned-agricultural-lands-help-save-the-planet#:~:text=The%20United%20States%20has%20lost,the%20start%20of%20the%20century.
12. G. Viglione, "What does the world's reliance on fertilisers mean for climate change?" Carbon Brief, 11 July 2022, https://www.carbonbrief.org/qa-what-does-the-worlds-reliance-on-fertilisers-mean-for-climate-change/.
13. G. Monbiot, *Regenesis: Feeding the World Without Devouring the Planet* (London: Allen Lane, 2022), 77.
14. Polyphenols are a group of naturally occurring chemicals, however those produced by conifers have a three-to-six-month allelopathic effect meaning that they have a survival mechanism which destroys nearby plants by inhibiting seed sprouting and root development.
15. B. Raskin, *The Woodchip Handbook* (White River Junction, VT: Chelsea Green, 2021), 114.
16. There would be emissions from the machinery used for chipping (although it could be green energy) plus some transport emissions connected to field spreading but distances would all be within the farm's geography.
17. Raskin, *Woodchip Handbook*, 114.
18. Valujeva *et al.*, "Abandoned farmland".
19. Monbiot, *Regenesis*, 109.
20. R. Brown, "Farmers voice frustration as UK land use strategy delayed to 2024", Business Green, 21 December 2023, https://www.businessgreen.com/news/4159167/farmers-voice-frustration-uk-land-strategy-delayed-2024.
21. Scottish Government, "Scotland's Third Land Use Strategy 2021–26",

24 March 2021, https://www.gov.scot/publications/scotlands-third-land-use-strategy-2021-2026-getting-best-land/.

22. In addition our rural areas provide water purification and nutrient recycling.
23. Monbiot, *Regenesis*, 77.
24. Valujeva *et al.*, "Abandoned farmland".
25. N.-G. Wunsch, "Food industry in the UK: statistics and facts". Statista, https://www.statista.com/topics/7229/food-industry-in-the-uk/.
26. J. Kang. & L. Zinnia, "Meet the high-tech urban farmer growing vegetables inside Hong Kong's skyscrapers", *Forbes*, 23 May 2022, https://www.forbes.com/sites/johnkang/2022/05/23/meet-the-high-tech-urban-farmer-growing-vegetables-inside-hong-kongs-skyscrapers/?sh=5e1e699d3294.
27. J. Mock, "What if we all ate a bit less meat?" *New York Times*, 21 August 2019, https://www.nytimes.com/2019/08/21/climate/what-if-we-all-ate-a-bit-less-meat.html.
28. Monbiot, *Regenesis*, 77.
29. The Woodland Trust & The National Sheep Association, "The role of trees in sheep farming", July 2018, https://www.woodlandtrust.org.uk/publications/2018/07/role-of-trees-in-sheep-farming/.
30. S. González, "Extensive grazing and habitats conservation", MAVA Foundation, October 2020, https://trashumanciaynaturaleza.org/wp-content/uploads/2021/03/TyN-2020-Extensive-grazing-and-HCI.pdf.
31. *Ibid.*
32. We could add a sixth – onshore wind – but while it will become increasingly part of the energy mix in the future, relatively speaking, compared to the four primary land uses outlined, it will be on a significantly smaller scale.

10 Harnessing the climate benefits of agroforestry

1. R. McKie, "Nearly half of Britain's biodiversity has gone since industrial revolution", *The Guardian*, 10 October 2021, https://www.theguardian.com/environment/2021/oct/10/nearly-half-of-britains-biodiversity-has-gone-since-industrial-revolution.
2. "Biodiversity", World Wildlife Fund, https://www.wwf.eu/what_we_do/biodiversity/.
3. Department for Environment, Food and Rural Affairs, "Agricultural land use in England", 29 September 2022, https://www.gov.uk/government/statistics/agricultural-land-use-in-england/agricultural-land-use-in-england-at-1-june-2022 and Eurostat, "Farms and farmland in the European Union: statistics".
4. "About agroforestry", Agroforestry Research Group, https://www.agroforestry.co.uk/about-agroforestry/.

5. European Commission, "Sustainable carbon cycles for a 2050 climate-neutral EU: technical assessment", Commission Staff Working Document, 15 December 2021, https://climate.ec.europa.eu/system/files/2021-12/swd_2021_451_parts_1_to_3_en_0.pdf.
6. "Agroforestry for Europe", AGFORWARD, https://www.agforward.eu/; "Trees for timber intercropped with cereals", https://www.agforward.eu/trees-for-timber-intercropped-with-cereals-445.html.
7. M. den Herder *et al.*, "Current extent and stratification of agroforestry in the European Union", *Agriculture, Ecosystems & Environment* 241 (2017): 121–32; https://www.sciencedirect.com/science/article/abs/pii/S0167880917301159.
8. R. Harrabin, "Tree planting 'can reduce flooding'", BBC News, 11 March 2016, https://www.bbc.co.uk/news/science-environment-35777927.
9. "A carbon neutral sheep, cattle and agroforestry farm in south-west Victoria, Australia", Jigsaw Farms, https://www.jigsawfarms.com.au/.
10. "Agroforestry", Soil Association, https://www.soilassociation.org/causes-campaigns/agroforestry/.
11. Briggs in conversation with the author.
12. "Biota" is the animal and plant life of a particular region, habitat, or geological period.
13. Briggs in conversation with the author.
14. "Biome" is a large naturally occurring community of flora and fauna occupying a major habitat, for example, forest or tundra.
15. P. Burgess *et al.*, *The Agroforestry Handbook*, The Soil Association, https://www.soilassociation.org/farmers-growers/low-input-farming-advice/agroforestry-on-your-farm/download-the-agroforestry-handbook/.
16. Farm Carbon Calculator, https://calculator.farmcarbontoolkit.org.uk/.
17. Agrecalc, https://www.agrecalc.com/.
18. P. Burgess & A. Graves, "The potential contribution of agroforestry to net zero objectives", Report for the Woodland Trust, Cranfield University, 1 November 2022, https://dspace.lib.cranfield.ac.uk/handle/1826/18664.
19. Email correspondence to author from Lawson, a policy analyst with the EURAF and a former research ecologist and Chartered Forester for the UK's Natural Environment Research Council.
20. "Improve livestock health and wellbeing", Woodland Trust, https://www.woodlandtrust.org.uk/plant-trees/david-brass/.
21. "Perennial energy crops as a source of food for bees?" Energy Crops Consultancy, https://www.energycropsconsultancy.com/biodiversity-with-src-willow/bees-and-pollinating-insects/.
22. C. Kennedy & T. Southwood, "The number of species of insects associated

with British trees: a re-analysis", *Journal of Animal Ecology* 53:2 (1984): 455–78.

23. C. Tosh & S. Westaway, "Incentives and disincentives to the adoption of agroforestry by UK farmers: a semi-quantitative evidence review", Organic Research Centre, Cirencester, January 2021, https://www.organicresearchcentre.com/wp-content/uploads/2021/06/AF-ELM-Test-Evidence-Review.pdf.
24. *Ibid.*
25. *Ibid.*
26. *Ibid.*
27. Briggs in correspondence with the author.
28. *Ibid.*
29. V. Milicevic, "The European Union and forests", European Parliament fact sheet, October 2023, https://www.europarl.europa.eu/factsheets/en/sheet/105/the-european-union-and-forests#:~:text=Taking%20the%20definition%20given%20above,world's%20total)%20in%20the%20EU; Eurostat, "Farms and farmland in the European Union: statistics". Calculation provided to the author by Gerry Lawson, Policy Analyst, European Agroforestry Federation.

11 Factory-built houses

1. HUF HAUS, https://www.huf-haus.com/en-uk/.
2. Agile Homes, https://agile.property/.
3. "Modern methods of construction", Go Construct, https://www.goconstruct.org/educational-resources/learn-about-construction/modern-methods-of-construction/#:~:text=Modern%20methods%20of%20construction%20is,quality%20homes%20in%20less%20time.
4. F. Barbosa *et al.*, "Reinventing construction: a route to higher productivity", McKinsey Global Institute, February 2017, https://www.mckinsey.com/~/media/McKinsey/Industries/Capital%20Projects%20and%20Infrastructure/Our%20Insights/Reinventing%20construction%20through%20a%20productivity%20revolution/MGI-Reinventing-construction-A-route-to-higher-productivity-Full-report.pdf.
5. M. Abdel-Wahab *et al.*, "Trends of skills and productivity in the UK construction industry", *Engineering, Construction and Architectural Management* 15:4 (2008): 372–82.
6. P. Brien & M. Ward, "Future of small and medium-sized housebuilders", House of Commons Library Research Briefing, 5 May 2023, https://commonslibrary.parliament.uk/research-briefings/cdp-2023-0100/.

7. "Greener, better, faster: Modular's role in solving the housing crisis", Make UK, https://www.makeuk.org/insights/reports/greener-better-faster-modulars-role-in-solving-the-housing-crisis.
8. White in correspondence with the author.
9. R. Wilson, "Building study: George Clarke and TDO's modular house for Urban Splash", *Architects' Journal*, 12 April 2018, https://www.architectsjournal.co.uk/buildings/building-study-george-clarke-and-tdos-modular-house-for-urban-splash.
10. S. Sharma, "Check out TV architect George Clarke's amazing space – in North Shields", *Chronicle Live*, 9 February 2018, https://www.chroniclelive.co.uk/news/north-east-news/george-clarke-smiths-dock-houses-14269208.
11. K. McVeigh, "Tyre dust: the 'stealth pollutant' that's becoming a huge threat to ocean life", *The Guardian*, 25 July 2022, https://www.theguardian.com/environment/2022/jul/25/tyre-dust-the-stealth-pollutant-becoming-a-huge-threat-to-ocean-life.
12. J. Wilmore, "Legal & General to 'cease production' of new homes at modular factory", *Inside Housing*, 4 May 2023, https://www.insidehousing.co.uk/news/legal--general-to-cease-production-of-new-homes-at-modular-factory-81335.
13. M. Lozanovska, "Prefab housing and the future of building: product to process", *Fabrications: The Journal of the Society of Architectural Historians* 29:2 (2019): 293–310.
14. Hairstans in conversation with the author.
15. It is Germany's goal to build 400,000 homes per year, not Nokera's. In the UK, 300,000 has been the target.
16. Nokera, https://nokera.com/en/.
17. TopHat, https://tophat.io/.
18. E. Twinch, "Barratt to open £45m timber frame factory", *Housing Today*, 8 March 2023, https://www.housingtoday.co.uk/news/barratt-to-open-45m-timber-frame-factory/5122111.article.
19. E. Twinch, "Persimmon joins £70m investment into modular housebuilder TopHat", *Building*, 25 April 2023, https://www.building.co.uk/news/persimmon-joins-70m-investment-into-modular-housebuilder-tophat/5122878.article?ref=ctvc.co.
20. *Ibid.*
21. BoKlok, https://www.boklok.co.uk/about-us/.
22. *Ibid.*

Conclusion: Timber Rising!

1. A. Niranjan, "'Era of global boiling has arrived,' says UN chief as July set to be hottest month on record", *The Guardian*, 27 July 2023, https://www.theguardian.com/science/2023/jul/27/scientists-july-world-hottest-month-record-climate-temperatures.
2. R. Pasternack *et al.*, "What is the impact of mass timber utilization on climate and forests?" *Sustainability* 14:2 (2022): 758.
3. Hemp, sheep's wool, mycelium, flax, etc.; see L. Mark, "Natural insulation materials", *Architects' Journal*, 21 February 2013, https://www.architectsjournal.co.uk/specification/natural-insulation-materials.
4. "Treelogy, Part II".
5. Potsdam Institute for Climate Impact Research, "Wood instead of steel and concrete: Schellnhuber calls for a green Bauhaus movement", 21 April 2021, https://www.pik-potsdam.de/en/news/latest-news/wood-instead-of-steel-and-concrete-schellnhuber-calls-for-a-green-bauhaus-movement.
6. IPCC, "Climate change and land: an IPCC special report on climate change, desertification, land degradation, sustainable land management, food security, and greenhouse gas fluxes in terrestrial ecosystems". https://www.ipcc.ch/srccl/download/.
7. *Ibid.*
8. F. Williams, "'Over 90% of concrete used in construction could be replaced with timber'". *Architects' Journal*, 20 July 2023, https://www.architectsjournal.co.uk/specification/in-practice-over-90-of-concrete-used-in-construction-could-be-replaced-with-timber.
9. Richard Stanford, chief executive, Forestry Commission, BBC Radio 4 Today Programme, 28 March 2023.
10. Calculation provided to author by Gerry Lawson, Policy Analyst, European Agroforestry Federation.
11. European Academies Science Advisory Council, "Negative emission technologies: what role in meeting Paris Agreement targets?" EASAC policy report 35, February 2018, https://unfccc.int/sites/default/files/resource/28_EASAC%20Report%20on%20Negative%20Emission%20Technologies.pdf and UK Climate Change Committee, "The sixth carbon budget: greenhouse gas removals", https://www.theccc.org.uk/wp-content/uploads/2020/12/Sector-summary-GHG-removals.pdf.
12. "UK urban canopy cover", Forest Research, https://www.forestresearch.gov.uk/research/i-tree-eco/uk-urban-canopy-cover/.
13. European Environment Agency, "How green are European cities? Green space key to well-being – but access varies", 1 February 2022, https://www.eea.europa.eu/highlights/how-green-are-european-cities.

14. "UK urban forest can store as much carbon as tropical rainforests", UCL News, 26 June 2018, https://www.ucl.ac.uk/news/2018/jun/uk-urban-forest-can-store-much-carbon-tropical-rainforests.
15. "Trees in our cities: 10 reasons we need to plant more", Trees for Cities, https://www.treesforcities.org/stories/trees-in-our-cities-10-reasons-we-need-to-plant-more.
16. Carbon offsets generated by the 10 per cent agroforestry would not be available as the farms themselves would need to "keep" the offsets to use to help enable their farm to reach net zero. Once a farm was net zero any additional CO_2 savings generated from tree planting could be sold.
17. Twinch, "Barratt to open £45m timber frame factory".
18. "Investing in the modern forest and timber construction industry", Timber Finance, white paper 2023, https://25003384.fs1.hubspotusercontent-eu1.net/hubfs/25003384/Whitepaper_EN.pdf.
19. Wood is used to provide the mould for concrete to be poured into and is then discarded. A large number of building materials are delivered on wooden pallets to construction sites and then discarded.
20. M. Kim & H. Song, "Analysis of the global warming potential for wood waste recycling systems", *Journal of Cleaner Production* 69 (2014): 199–207.
21. "Natural insulation: what do you need to know?" Critical Concrete, 30 January 2023, https://criticalconcrete.com/natural-insulation-what-do-you-need-to-know/.
22. Climate Change Committee, "Progress in reducing emissions: 2023 Report to Parliament", https://www.theccc.org.uk/wp-content/uploads/2023/06/Progress-in-reducing-UK-emissions-2023-Report-to-Parliament-1.pdf (agriculture 11 per cent and shipping 3 per cent).
23. European Environment Agency, "Total greenhouse gas emissions by sector (%) in EU-27, 2009", https://www.eea.europa.eu/data-and-maps/figures/total-greenhouse-gas-emissions-by-sector-in-eu-1 (agriculture 8.6 per cent, international shipping 3.2 per cent, and aviation 3.1 per cent).
24. DEFRA, *Timber in Construction Roadmap*, UK Government, 2023, https://www.gov.uk/government/publications/timber-in-construction-roadmap/timber-in-construction-roadmap.
25. "Engineered timber: we're expanding our underwriting appetite", Aviva, 2 November 2023, https://gcs.aviva.com/en-gb/news/engineered-timber/.
26. L. Crook, "Twenty per cent of new homes in Amsterdam to be constructed from timber", De Zeen, 2 November 2021, https://www.dezeen.com/2021/11/02/amsterdam-new-buildings-20-per-cent-timber/.
27. This fact sourced from the Director of Swedish Wood Anna Ryberg Agren in correspondence with the author. "Mid-rise" is the term usually used for

buildings that are between four and 12 storeys high and if modern they will have an elevator. They provide the majority of housing in city centres around the world. "High-rise" is usually the term used for anything beyond 12 storeys.

28. "Sweden wants to build an entire city from wood", *The Economist*, 21 June 2023, https://www.economist.com/science-and-technology/2023/06/21/sweden-wants-to-build-an-entire-city-from-wood.
29. RBC Climate Action Institute, "Timber rising: how wood can spur Canada's green building drive", https://thoughtleadership.rbc.com/wp-content/uploads/Buildings_report.pdf.
30. Council on Tall Buildings and Urban Habitat, "The state of tall timber: a global audit", February 2022, https://www.ctbuh.org/mass-timber-data.
31. Mass Timber Strategy, "The Marshall effect: what it is and why it matters", https://www.masstimberstrategy.com/the-marshall-effect.
32. Forest and Climate Leaders' Partnership, "Initiative for greening construction with sustainable wood", press release, 6 December 2023, https://forestclimateleaders.org/wp-content/uploads/2023/12/FCLP-COP28-Release-Buildings-06122023.pdf.
33. Potsdam Institute for Climate Impact Research, "Wood instead of steel and concrete".
34. A. Himes & G. Busby, "Wood buildings as a climate solution", *Developments in the Built Environment* 4 (2020), https://www.sciencedirect.com/science/article/pii/S2666165920300260.
35. R. Solnit, *Hope in the Darkness: The Untold History of People Power* (Edinburgh: Canongate, 2005), 5.

Index

Note: Page numbers in *italics* and **bold** denote figures and tables, respectively.